Language and Complex Systems

An understanding of language as a complex system helps us to think differently about linguistics, and helps us to address the impact of linguistic interaction. This book demonstrates how the science of complex systems changes every area of linguistics: how to make a grammar, how to think about the history of language, how language works in the brain, and how it works in social settings. Kretzschmar argues that to construct the best grammars of languages, it is necessary to understand the complex system of speech. Each chapter makes specific recommendations for how linguists should manage empirical data in order to form better generalizations about a language and its varieties. The book will be welcomed by students and scholars working in linguistics and English language, especially the study of language variation and the historical development of English.

WILLIAM A. KRETZSCHMAR, JR. teaches English as Harry and Jane Willson Professor in Humanities at the University of Georgia. He also has appointments at the University of Glasgow in Scotland and the University of Oulu in Finland. His recent publications include *The Linguistics of Speech* (Cambridge, 2009).

Language and Complex Systems

William A. Kretzschmar, Jr.

CAMBRIDGE
UNIVERSITY PRESS

CAMBRIDGE
UNIVERSITY PRESS

University Printing House, Cambridge CB2 8BS, United Kingdom

One Liberty Plaza, 20th Floor, New York, NY 10006, USA

477 Williamstown Road, Port Melbourne, VIC 3207, Australia

4843/24, 2nd Floor, Ansari Road, Daryaganj, Delhi - 110002, India

79 Anson Road, #06-04/06, Singapore 079906

Cambridge University Press is part of the University of Cambridge.

It furthers the University's mission by disseminating knowledge in the pursuit of education, learning and research at the highest international levels of excellence.

www.cambridge.org
Information on this title: www.cambridge.org/9781107100459

First published 2015

A catalogue record for this publication is available from the British Library

ISBN 978-1-107-10045-9 Hardback

Turning and turning in the widening gyre
The falcon cannot hear the falconer;
Things fall apart; the centre cannot hold;
Mere anarchy is loosed upon the world,
The blood-dimmed tide is loosed, and everywhere
The ceremony of innocence is drowned;
The best lack all conviction, while the worst
Are full of passionate intensity.

Surely some revelation is at hand;
Surely the Second Coming is at hand.
The Second Coming! Hardly are those words out
When a vast image out of *Spiritus Mundi*
Troubles my sight: somewhere in sands of the desert
A shape with lion body and the head of a man,
A gaze blank and pitiless as the sun,
Is moving its slow thighs, while all about it
Reel shadows of the indignant desert birds.
The darkness drops again; but now I know
That twenty centuries of stony sleep
Were vexed to nightmare by a rocking cradle,
And what rough beast, its hour come round at last,
Slouches towards Bethlehem to be born?

The Second Coming (1920), William Butler Yeats

Contents

Figures

Tables

Acknowledgements

This book, the sequel to my 2009 book *The Linguistics of Speech*, began before the first book was even published. Chapter 1 began as a talk at a 2008 conference organized by Manfred Krug, himself one of the earliest to write about string frequency. While *The Linguistics of Speech* laid the foundations for the study of speech as a complex system, the chapters in this book apply complex systems to many fields within linguistics. I am grateful to the audiences of the many papers on applications in different fields that I prepared and improved over time. These audiences, some at professional meetings and many at universities where I was invited to speak, were always enthusiastic, always full of questions about how complexity science might help them to understand better the linguistic phenomena in which they were particularly interested.

I would also like to thank Jane Willson, whose generosity to the University of Georgia created the professorship I hold. Resources from the professorship allowed me to keep up a busy schedule of presentations in order to bring the idea of complex systems to new audiences of linguists and people in digital humanities. Without the kind of assistance provided by Jane and her late husband Harry, using resources from their Sunnyland Farms business, new ideas would have a much harder time getting out to be considered in the marketplace of ideas. They were thus benefactors not only of the university but of science more generally. I will continue to give pecans from Sunnyland Farms to friends, and I remain proud, Jane, to be your nutty professor.

Finally, I would like to thank my predecessors on the Linguistic Atlas Project for their friendship and support over many years. Raven and Virginia McDavid brought me into that fold when I was just a graduate student, and Lee Pederson has shared ideas with me for nearly thirty years. Without the Atlas, of course, I would be unlikely to have seen the distributional properties that prove speech to be a complex system. Pederson and the McDavids kept Atlas survey research alive while many others pursued sociolinguistics instead, and so preserved the chance for an independent viewpoint.

WAK

Introduction

Cause and effect. Action and reaction. The principle is built into the Western tradition, both in science and in the popular imagination. It is an expectation for the way things work, for the way things must work. If you push the first domino then it will fall, and the force will be transmitted to all of the other dominoes in as long a sequence as you might care to set up. We commonly extend the idea to domains beyond the physical, as for instance the "domino theory" according to which allowing one state to change its politics will in turn cause other states to do so as well. How could we ever have an effect without a cause? When we see that something has happened and the cause is not immediately apparent, we expect that we can find the cause, or we believe that somebody else sufficiently motivated and sufficiently skilled could eventually find the cause if they tried hard enough. In the domain of language, we have been used to thinking in generative terms that a universal grammar is the cause of human language, presumably a grammar with biological connections of some sort because language is a property of the species that no other living species shares in quite the same way or to the same extent. We have expected that somehow, someday, somebody would find that cause, would reveal universal grammar for what it really is.

This book takes a different position, not just about universal grammar and language but about cause and effect itself. Surely the principle of cause and effect does operate in a great many domains. However, in recent decades it has become clear that some effects arise without particular causes, as the result of random interactions of large numbers of elements in complex systems. Such effects just *emerge* from the random interactions, without any deterministic cause; given the same elements, different effects can and do emerge. Adding grains of sand to a pile makes a larger and larger pile until landslides begin to occur, and mostly small and some large landslides then continue to occur unpredictably as grains continue to be added to the pile (Kauffman 1995: 28–29). There is no simple cause and effect relationship between the sand, the pile, and the size of the landslides. Landslides thus emerge from the system: we can predict that landslides will occur, but we cannot predict the exact timing or size or location of any individual landslide. Stuart Kaufmann has called this kind of emergence "order for free" (1995: 83), by which he means that we achieve regularity in behavior, the certainty that there will be larger and smaller landslides as more sand is added to the pile, without any simple cause or assemblage of simple causes. We cannot reduce the landslide problem to cause and effect, and so we get "order for free" in the sense that the regularities do not occur at the cost of particular causes. Speech, language in use, is like landslides: order in language just emerges from the linguistic interactions of speakers, agents using speech. I demonstrated this fact from first principles in an earlier book, *The Linguistics of Speech* (2009), which also deals with the issue that speakers as agents are much, much more complex and hard to describe than grains of sand falling on a pile. In this book, I will show how an understanding of language in use as a complex system helps us to think differently about a number of problems in linguistics, and helps us to address the great complexity of linguistic interactions.

In the chapters that follow, I will first introduce the central ideas of complexity theory as they are relevant to human culture, and especially to speech, and then in each subsequent chapter treat an important issue in linguistics to which an understanding of complex systems can bring new perspective. These chapters do not offer a replacement for well-established and successful methods of structural and generative linguistics, derived as these are from a different, causal scientific approach. These traditional methods arose in an attempt to bring new science to problems in the study of language, along with new social sciences that sought to do the same thing in other realms of human experience. The social sciences have not failed (*pace* Wilson and Gould, as described in Chapter 2) but rather have made great progress over the last century in our understanding of human social issues, from cultural systems in anthropology to cognition in psychology. A new humanities approach to the study of language with complex systems adds the same kind of alternative perspective that complex systems add to reductionism in other sciences, from physics to evolutionary biology to economics, at the same time that it respects the social history of the different disciplines. Every humanist knows that one's point of view means a great deal to the understanding of whatever facts and problems we wish to confront.

That said, even though the methods of studying complex systems do not replace those of generative and structural linguistics, no linguist can afford to ignore the fact that human language is a complex system. All approaches to human language must begin with speech, and all speech is embedded in the complex system. We all must take account, first, of speech as people use it. This means that we cannot remain satisfied with traditional assumptions about grammaticality, in which we rely on speakers' intuitions about what pronunciations, words, and constructions belong to their languages and which ones do not. The complex systems view tells us that 80 percent of what people can and do say occurs only rarely, so that most speakers would not use those forms.

We have traditionally considered the remaining common forms to be "grammatical," and yet the rare forms are no less a part of one's language. We all use some of the rare forms along with the common ones, just different uncommon forms for different speakers. Human language has a great capacity to allow for and retain a wide range of alternate possibilities for how to say the same thing. Moreover, the scaling property of complex systems tells us that other people are bound to vary in how often they use any of the possibilities, in groups at every level of size. While each group of speakers, from neighborhoods and small communities of practice all the way up to regional, national, and super-national groups, will use a great many of the same rare and common variants, we can always distinguish the usage of different groups by the different frequencies with which the speakers use the different variants. Our own experience as an individual speaker, which includes our membership in the large number of different groups to which we each belong, thus provides poor evidence for what may be "grammatical" for others. We all use a language ourselves at our own nexus of spatial and social and textual possibilities in the wider society around us. Our own intuitions about our own language give us little sense of the range of alternative possibilities available in the wider range of speakers of our language, and we have only a little more sense about the frequencies with which speakers from different groups use the available variants. The most basic assumption of generative and structural linguistics, that we speakers all share the system of a language, share the rules for a language, is simply wrong. We all participate in speech, but the language is a little different, both in its available features and in the frequency with which we use those features, for each one of us individually and for each of us as a participant in every group to which we belong. When we think about rules and systems, these are actually generalizations that we make after the fact from our perceptions of our own language and the language we observe around us; they are not generative or structural in any sense essential to language, but instead just serve to help us organize our perceptions.

The only way to cope with these distributional facts is to take account of them from the beginning. If we do so, we can still make grammars as generalizations about the usage of different groups of speakers, now with better knowledge of how the complex system of language in use forms the empirical basis for those generalizations. Moreover, special areas of linguistics like usage-based linguistics, historical linguistics, cognitive linguistics, and sociolinguistics have all brought traditional assumptions of grammaticality to their own pursuits. The complex systems view can isolate problems in these areas, and a knowledge of complex systems makes it possible to see opportunities for improvements. Therefore, in this volume I will review some of the key points and the literature in these areas, identify problems, and suggest possibilities for a new way forward.

The realization that speech is a complex system is a new and fundamental fact about language. It is not the frosting on the cake, but provides us with an understanding of the flour out of which the cake is made. The realization that speech is a complex system changes everything for linguists. Yes, we can still make grammars, we can still study usage, history, cognition, and social aspects of language, but now we must do so with a new idea about the empirical basis for the generalizations we want to make. This kind of change has happened in other fields. The idea that earth, air, fire, and water were the basic elements of the natural world had to be replaced with the notion of atoms. The idea that the earth was at the center of the universe had to be replaced with the notion of the sun at the center of our solar system. The assumption of Darwinian gradual evolution has more recently had to be replaced with the notion of punctuated equilibrium. It is no overstatement to say that linguists now have to make a similar transition. Our paradigm for how to think about language has changed. Our most basic evidence about language has turned out to be different from what we had earlier expected and assumed. This book suggests why we should and how we can face that fact.

1 Language and complex systems

In Mitchell's (2009: 13) definition, a complex system is "a system in which large networks of components with no central control and simple rules of operation give rise to complex collective behavior, sophisticated information processing, and adaptation via learning or evolution." Complexity, in this sense, does not just mean 'complicated.' The etymology of the word refers to Latin *complexus*, and thus to a number of things encompassed together; something that is *complex* has acquired in English the more specialized meaning 'interwoven,' which conforms to the etymological origins of *complicated*. We can talk about a *complex problem* or a *complex relationship* without meaning anything much more than that the problem or the relationship has many parts that are intertwined together. But Mitchell means something much more specialized that that.

The scientific field known as complexity science can trace its history back to 1984, when the Santa Fe Institute (SFI) was founded (for more details and the quotations here, formerly in the Profile/History statement on the SFI website, see www.intelros.ru/subject/karta_bud/10841-santa-fe-institute.html). Complex systems, also called complex adaptive systems, are "firmly grounded in the quantitative methods of physics, chemistry, and biology" as they are studied at the Institute, and work there extends to a wide range of topics including "global climate, financial markets, ecosystems, the immune system, and human culture." SFI was originally the idea of scientists at the Los Alamos National Laboratory like the chemist George Cowan, and attracted support from well-known physicists Murray Gell-Mann and David Pines and mathematician Gian-Carlo Rota. Also significant was support from the CEO of Citibank, John Reed, who saw complex systems as an alternative to traditional economics. SFI remains focused on a new approach to science:

> doing complexity science required a commitment to observation, experience, and experiment that is in balance with the transdisciplinary, out-of-the-box curiosity that gave rise to the original question. It's not that outcomes in complex adaptive systems are repeatable as they are in many scientific disciplines; complex systems are by definition unpredictable, and often downright squirrelly. But finding the patterns embedded in complex systems requires a distinct brand of scientific rigor and methodological approaches that in many cases haven't yet been invented.

Complexity science, then, grew up at SFI and elsewhere in opposition to what the founders thought was the "stove-piped, bureaucratic, funding-centric approach to science that had taken hold both at the federal laboratories and academia." At bottom, modern complexity science is committed "to looking at any problem through an empirical lens and seeking quantitative patterns" in its founding scientific fields but also in economics, the other social sciences, and the humanities. Complex systems, as Mitchell defines them, stand in opposition not just to the funding patterns of modern science but also as to cause-and-effect reductionism, the model most often applied in modern science. Complexity science does not abandon the empirical observation, rigorous methods, and quantitative analysis that characterize modern science, but also does not expect that simple causes can be found for the effects we observe in "large networks of components with no central control," the domain where complex systems can be found. Complexity science does not replace reductionism completely, which works very well in many domains, but rather serves as a better alternative for a scientific model in its own domains. As I argued in Kretzschmar (2009), speech is one of those domains.

While complexity science was taking off at SFI, it did receive some early allusive discussion in linguistics: Lindblom, MacNeilage, and Studdert-Kennedy published a 1984 paper on self-organizing processes in phonology; Paul Hopper presented his seminal paper called "Emergent Grammar" in Berkeley in 1987 (see Chapters 3 and 4); Ronald Langacker published a chapter titled "A Usage-Based Model" for cognitive linguistics in 1988. Gradually more papers attempting to use complex systems in linguistics appeared in the 1990s, such as Van Geert (1991). In 1996 Edgar Schneider presented a paper whose title was a question, "Chaos Theory as a Model for Dialect Variability and Change?" (published 1997). At that time, it had already been over twenty years since the original paper on climate by Edward Lorenz that asked the question, "Does the Flap of a Butterfly's Wings in Brazil Set off a Tornado in Texas?" (1972), and over ten years since the founding of the SFI, where chaos theory was studied as part of the emerging field of complexity science. But it was very early for a student of language to consider the subject as a serious model for speech. In the same year, J. K. Chambers commented in a book review that "We will need a coterie of sociolinguists expert in chaos theory before we can make a start [at applications to our field]" (1996: 163). Chambers noted that the biggest problem for language applications then was that chaos theory seemed to require a long series of observations over time, a rare commodity for those who systematically record language in use. At about the same time, Diane Larsen-Freeman (1997) suggested the use of complexity science for the study of language acquisition. Five years later, Joan Bybee proposed that frequency effects observable in language use could be connected with complex systems, in that linguistic structure itself might thereby be generated (2001, discussed more fully in Chapter 3). In the meantime, there was no thoroughgoing attempt to align evidence from speech with the basic principles of complexity science until Kretzschmar (2009).[1] As will become clear, not only does language in use conform to the requirements of a complex system, analysis of speech evidence in terms of a complex system can help to resolve a number of long-standing problems faced by linguists.

Examples of complex systems

Let us begin with a simple and familiar example of a complex system: ants.[2] Ants are not smart. They can only do a few things, like exploit food sources, build nests, and defend themselves against intrusions, but no commissar of ants tells them to do any one of them. Instead, ants just happen to be doing one of their tasks at any given time. In searching for food, for example, ants wander around randomly; if they find some food, they leave chemical traces, called pheromones, along their path back to the colony (see Wilson (1998: 74–77) for a brief treatment of his discovery of ant pheromones). Other ants can follow the traces, and leave more traces on their way back, till the path becomes a line with lots of ants on it to exploit the food resource (Figure 1.1). However, not all of the ants in the nest follow the path; some keep foraging and some stay home on nest and defense duty, again not because they are told to do so by any central authority: the queen is the biological center of an ant colony but does not direct its activities. When ants come boiling out of a nest when it is disturbed, a great number of them are changing their behavior at once from food or nest duty to defense. The defense reaction of the ants is a local example of Gould's "contingency," the fact that current conditions, whether a meteor strike on the evolutionary scale, or an incautious human footstep for the ants, must influence the outcome of a complex system (see Chapter 2). However, not all of the ants leave to gather food or rush to defense given the stimulus to do so. Some stay on nest duty during a provocation, and some ants look randomly for food when most have joined the line to a known food source. What looks

[1] Kuhl (2003) does make a serious effort to align complexity science with particular data, but not for populations. Kuhl treats particular speaker interactions and, as Kuhl put it, "the idiolect as autonomous adaptive organism" (2003: 78). Dahl (2004) treats "linguistic complexity" in a different sense, from a highly formal point of view, in line with the general definition of complexity and not the more specialized sense presented here. Kretzschmar 2009 offers references to other linguists who make some allusion to complexity science, notably Mufwene's analogy of language of biological evolution (2008), or who develop a single aspect of it, notably Zipf's Law. More recently, Larsen-Freeman and Cameron (2008) and a group including Joan Bybee and Nick Ellis (Beckner et al. 2009) have discussed language structure as the outcome of a complex adaptive system, and these discussions will be treated in Chapter 3. Chapter 1 is based on Kretzschmar (2010).

[2] The discussion of ant behavior is derived from Mitchell (2009: 176–184).

Figure 1.1 Ants in a line. "Ant Trail." Author: Reyham Soenasto (www.myfolio.com © Reyham Soenasto; licensed under the Creative Commons Attribution-ShareAlike 2.0 Generic License)

to us like highly organized behavior is not controlled by any leader, and it is not absolutely determined by particular stimuli, but instead patterns of activity emerge from the random instinctual behaviors of ants as they are conditioned by circumstances. These patterns of activity, the result of the complex system, make the whole more than just the sum of a few instinctive behaviors.

From the point of view of a single ant, it merely does one of three things: finds and carries food, builds the nest, or defends the colony. Lacking any sort of stimulus, the behavior of different ants is random. What the ant does at any given time is influenced, but not determined, by what happens near it; when it detects chemical traces or touches antennae, an ant becomes more likely to enact one out of the three behaviors. Individual ants are subject to feedback from other ants. In a primitive way, they exchange information by means of their antennae and pheromones. So, if we now consider whole colonies of ants, we see that the feedback within an otherwise random system is what leads to development of frequency-based patterns of activity, which is what makes such a system complex. For an ant colony to survive, the system has to be random at its core and not deterministic. If all of the ants were deterministically required to follow the line to and from the food source, the nest could decay and might be lost to attack. If all of the ants just built the nest or stayed in defense mode, they would starve. Some of the ants need to continue enacting each of the behaviors.

On the other hand, feedback, the exchange of information even between creatures with no understanding, allows for the balance of activity, the frequency with which ants enact one of their behaviors, to shift towards the task that present circumstances demand. The contingencies of the moment do not cause just a small deviation from randomness: lots more ants are in the food line when that is the current need, and lots of ants rush out in defense, dangerously so for your foot on a Southern American fire ant mound, so the frequency profile of ant behavior is subject to sudden large changes. These changes in frequency profiles "emerge" as the current output of the complex system. On the evolutionary scale, Gould described such sudden large changes as "punctuated equilibrium" in the evolution of species (see Chapters 2 and 5). On the scale of ants, it may take only a few moments for defense mode to shift back to repair of the nest. Emergence, therefore, occurs continuously for complex systems. We often recognize it as having the shape of some structure, like a new species in evolutionary time or a line of ants leading to food, but emergence is founded upon frequency shifts in the behavior of the components in the complex system: ants, antibodies, genes, quanta in physics, agents making economic transactions, or innumerable other participants in complex systems. Finally, we can see that the complex system of the colony depends upon the continual motion of the ants. Each ant has to keep moving in order to allow the exchange of information that permits the colony as a whole to react to its current circumstances. It would literally be fatal for the ants to reach some equilibrium state and stop reacting to conditions. The real story of evolution is the same: adaptability over time in response to changing conditions, over a much longer timescale than the ant colony enjoys. In an era of climate change we may soon see what parts of our ecosystem and human cultural systems are adaptable enough to change with the climate as these complex systems react to changing conditions.

Ants are not smart, but they are animate. As it happens, animate life is not required for this kind of complex behavior to manifest itself. So-called "glider guns," from a computer program called Conway's Game of Life (Gardner 1970; animated versions can be found online, at Youtube or Wikipedia for instance; popular information or common knowledge will occasionally be cited from Wikipedia here, as a supplement to the usual citation of more authoritative sources), continually generate patterns that over time move across a matrix of squares that can be either white ("alive") or black ("dead"). The complex system of change in action in which the changing patterns in one part of the matrix continually generate "gliders" results from some simple rules as they

Figure 1.2 A board showing "gliders" (five cell combinations that begin in the middle and move to the lower right) from Conway's Game of Life (Gosper Glider Gun.YouTube video by nbidyanta.www.youtube.com/watch?v=pvUiA-Q-3hM)

are applied at any time step in the program: if any dead square is next to exactly three live squares, it becomes alive; if any live cell is next to two or three other live cells, it stays alive; in any other condition, the cell either stays dead or becomes dead. These simple rules generate changes in the pattern of live and dead cells across successive time steps. The initial state of the matrix clearly matters: if there were only a few scattered live squares to begin with, the matrix would immediately go dark; most arrangements of live cells do not produce the regular behavior of the gliders; but some arrangements of live cells do create the circumstances that, with these simple rules, generate a complex system. Again, for the complex system to work, it has to keep moving across time steps – we could not watch the pattern of gliders otherwise. The rules for the Game of Life represent the exchange of information in the program; as for the ants, given current circumstances, the rules allow the behavior of a cell to be influenced by neighboring cells between time steps.

The Game of Life is a particular instance of a mathematical model, most often implemented on a computer, called a cellular automaton. Life and death in cellular automata are just metaphors; the importance of discussing cellular automata here is to show that complex systems are amenable to mathematical analysis. Ants are hard to count, and of course the fact that ants can only do one of a few behaviors at a time simplifies life in the ant colony so that we can make better sense of it. Cellular automata offer us a much more controlled environment in which to study the operation of complex systems. It is possible, for instance, to study all possible outcomes of the simplest cellular automata as Stephen Wolfram has done (2002). Wolfram found that only a very small number of the possible outcomes are complex, as opposed to those that soon settle into a fixed or alternating state, or never settle down and so continue to be chaotic, essentially unpredictable except at very long timescales. It is also possible, not just to look at all the possible outcomes, but to develop mathematical models to describe the operation and outcomes of complex systems and then apply them to speech, as illustrated in Kretzschmar, Juuso, and Bailey 2014. Complex systems may not be rule-bound in any deterministic sense, but they are indeed describable in quantitative terms.

Principles of complex systems

We can sum up the basic operation of our examples of complex systems in just a few principles: (1) continuing dynamic activity in the system, (2) random interaction of large numbers of components, (3) exchange of information with feedback, (4) reinforcement of behaviors, (5) emergence of stable patterns without central control. Complex systems like this were originally described in the physical and biological sciences, and the definite procedures of information exchange have been explored in computer science (e.g. Holland 1998). Ilya Prigogine won the Nobel Prize in Chemistry for his discussion of complexity (see Nicolis and Prigogine 1977); Stephen Hawking includes complex systems in his latest popular book about physics, *The Grand Design* (Hawking and Mlodinow 2010); and the branch of economics that deals with complex systems is sometimes called "behavioral economics" or "complexity economics." These applications of complexity science explain aspects of the natural world, like ants or quantum mechanics, that appear not to conform to Newtonian calculus and traditional cause-and-effect explanations in science. The mathematician Benoit Mandelbrot claimed that "many patterns of Nature are so irregular and fragmented, that, compared to [standard geometry] Nature exhibits not simply a higher degree but an altogether different level of complexity" (1982: 1). His treatment of natural forms like the geometry of coastlines presented problems that could not be solved with traditional methods but required a new nonlinear mathematics, what he called "fractals." Fractals are familiar to many of us through repeating graphic designs like the "Koch islands" in Figure 1.3. The basic properties of fractals – including scaling properties, as illustrated here – characterize many objects of study in the physical, natural, and social sciences, not just graphic designs.

Complex systems, by definition, are open and not at equilibrium. As we learn from Stuart Kauffman,[3] we should begin with a distinction between a system at equilibrium and a non-equilibrium system. Equilibrium systems are closed; they do not exchange matter or energy outside the system, and the components they have are balanced. Equilibrium systems at low energy are just fixed arrangements of components. Kauffman offers the example of dropping a ball down the side of a bowl: the ball will roll up and down the sides but will

[3] This section is derived from Kauffman (1995: 10–20).

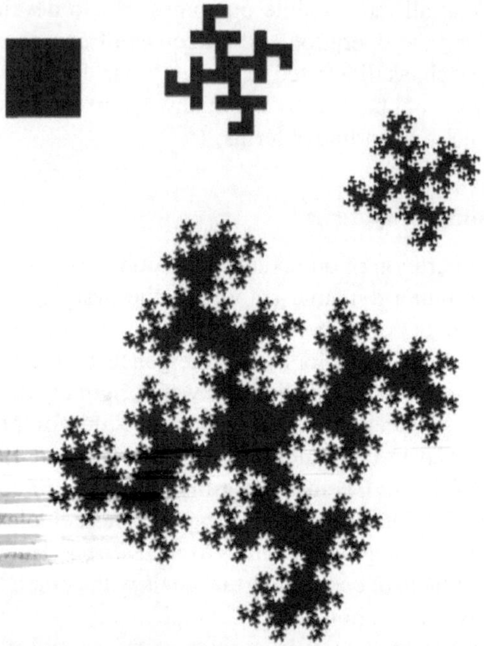

Figure 1.3 Koch island, at different scales (adapted from Mandelbrot 1982: 53)

eventually come to rest at the bottom of the bowl, at low-energy equilibrium. In the case of an energetic equilibrium system, again in Kauffman's example, if we put a quantity of gas molecules into a tank, the molecules do not stay ordered in a group at the point of entry; they keep moving around in the tank, and according to the ergodic theory (related to the statistical mechanics invented by Ludwig Boltzmann) they move randomly through all of the statistically possible states of arrangement. Order can be defined as the small number of possible states with the molecules clustered together, say in a corner of the tank. Or perhaps the molecules could be arranged in a regular matrix, equidistant from each other. While those states will in theory occur from time to time, their number is vanishingly small in comparison to all of the possible random states in which the molecules appear, more or less, to fill the tank. Thus, in an energetic equilibrium system like the gas molecules in a tank, order appears to vanish beneath the random movement of the gas molecules.

Against the gas molecules closed in a tank, Kauffman sets the counter example of the small whirlpool that forms near drains: this ordered structure will be maintained as long as the drain remains open and water continues to flow

Figure 1.4 Whirlpool as self-organization ("Whirlpool in the 'Water Garden' within Alnwick Garden." Author: Christine Westerback (geograph.org.uk. © Christine Westerback; licensed under the Creative Commons Attribution-ShareAlike 2.0 Generic License)

(Figure 1.4). The order in such a non-equilibrium system is sustained by persistent dissipation of matter and energy, and thus the whirlpool can be called a "dissipative structure" of the kind described by Prigogine. No stirring is required to start the whirlpool, no single and simple cause. The water is subject to natural laws like gravity that makes the water drain, but no collection of laws completely explains the whirlpool because randomness in the molecules and conditions is involved in the emergence of every particular whirlpool. While every whirlpool is a bit different from moment to moment and from day to day, the whirlpool always appears as an ordered state produced through the principle of self-organization in non-equilibrium systems. We can see the same thing in the eddies of a river, always there but ever-changing (Figure 1.5). The falling water of an old mill race (in Bamberg, Germany, a world heritage site) provides energy for a dissipative system in which these rafts of bubbles always emerge in changing patterns. The bubbles are never just distributed randomly like gas molecules in a tank. As for higher-level systems, Stuart Kauffman tells

Figure 1.5 Eddies and bubbles in the Regnitz

us that biological cells and the free-living systems composed of them are also complex systems which display order from self-organization.

Complex systems also must contain a large number of interactive components, whether the components are ants, water molecules, bubbles, or cells. Or, as Brian Arthur suggests, "economic agents [like] banks, consumers, firms, or investors" (1999: 107). The agents handle economic transactions, which are the actual components in the economic system. Just as Mandelbrot described nonlinear fractal regularities at different scales, in economics random interactions in complex systems lead to the emergence of non-random distributions without immediately apparent cause. So, let us consider an economic example, the historical demographic pattern for European markets, as described by J. C. Russell (1972: 29–37; Figures 1.6 and 1.7). Village markets, as in the Leicester locality, deal in a limited variety of chiefly local goods. Town markets deal in more specialized goods, and are spaced within a day's return travel of numerous villages. Cities are spaced at a still greater distance from numerous towns,

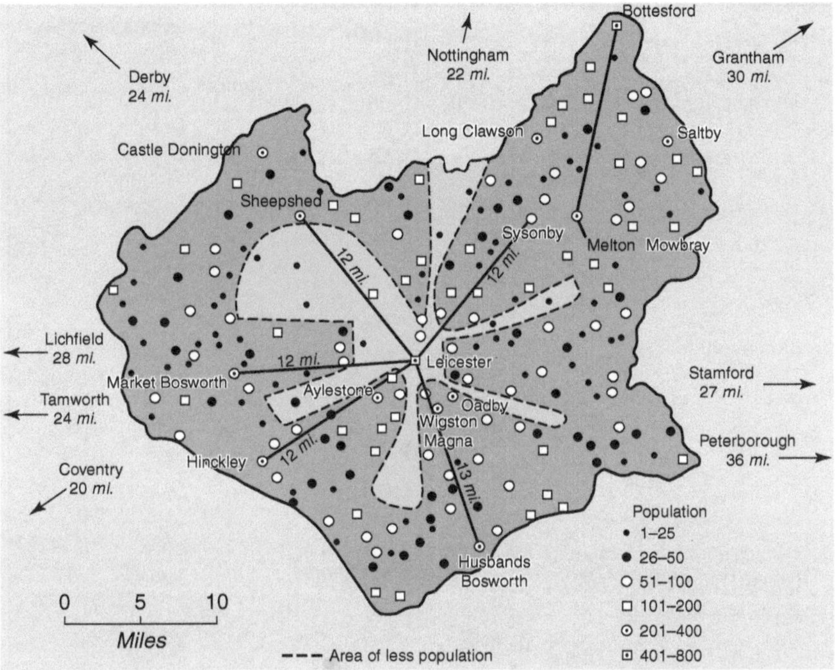

Figure 1.6 Leicester locality ca. 1086 (adapted from Russell 1972: 32)

as in the Barcelona region, and deal in still more specialized goods. Scaling is thus evident in the complex network of transactions, based on very large numbers of people interacting in markets, out of which a nonlinear distribution of markets emerges (many villages with few transactions, few cities with many transactions). Such scaling did not disappear with the passage of time. In contemporary industrialized countries we can buy essential items right in our local neighborhoods (in America often at a "convenience store"), but we can also go a bit further away and participate in the town market to buy things at a large grocery store, or at a general-purpose "big box store" (e.g. Walmart) or one that may specialize, say, in electronics or products for pets. If what we need is more specialized or exotic than what we can get in the town – like a favorite imported mustard, or somebody to repair a decades-old audio recorder – we may have to go to the big city, which supports markets in exotic foods and special services as well as having the resources available in neighborhoods and towns. This order at different scales, sometimes called a scale-free network, emerges not according to any specific plan or cause but as something that just

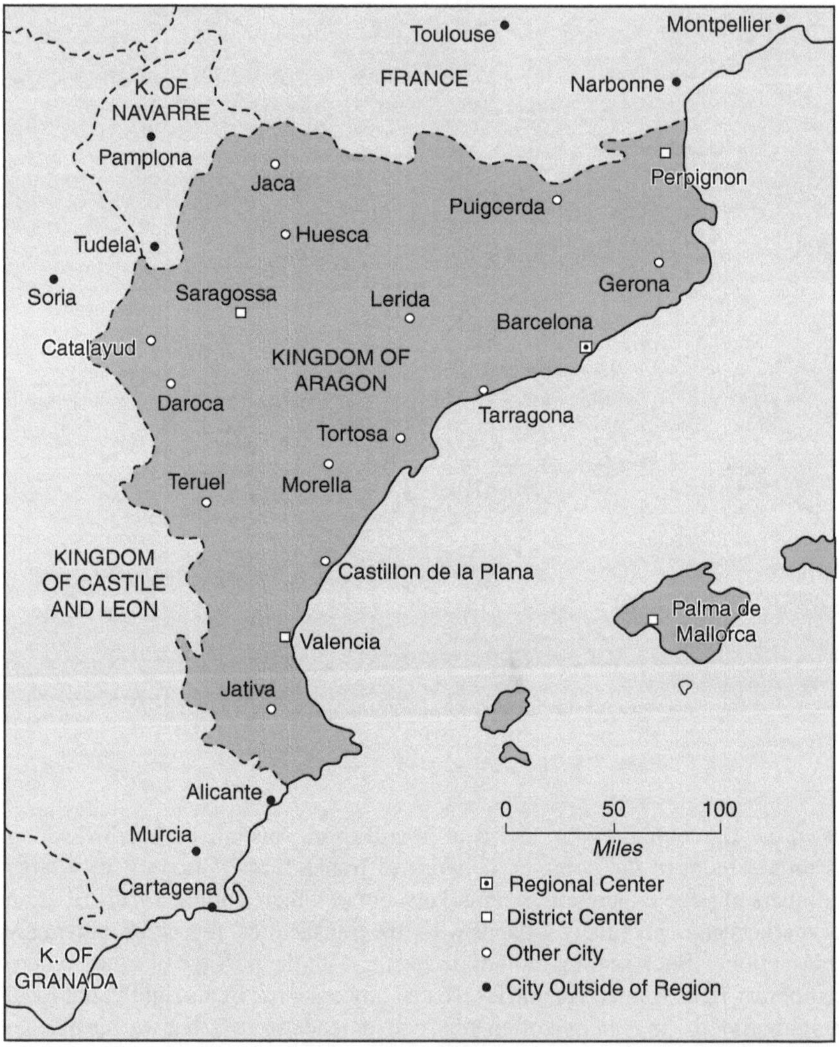

Figure 1.7 Barcelona region ca. 1360 (adapted from Russell 1972: 33)

happens, a result of chance operation in the complex system of the economic market.

As against these examples of complex systems, we can contrast the situation of chaotic systems, in which Schneider and Chambers were interested. Chaotic systems, not to be confused with the popular sense of *chaos* as 'disorder,' are said to be deterministic, since small changes in initial conditions lead to

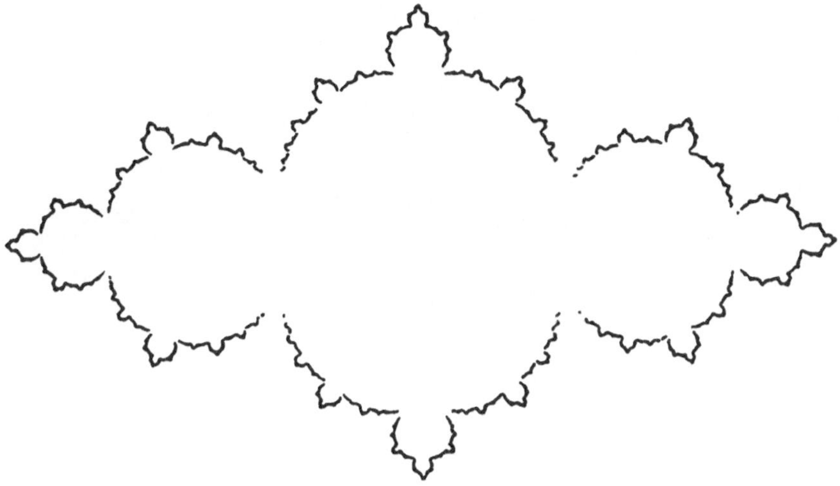

Figure 1.8 "Chaotic" fractal: San Marco Dragon (adapted from Mandelbrot 1982: 198)

significantly different future behavior, the famous butterfly effect. Complex systems do require sufficient conditions for their operations, such as enough live cells in the Game of Life to allow the rules to operate in a complex way, but they are not deterministic. So, for example, we still get a whirlpool or bubble patterns at different water levels, or if someone should wade in the water and disturb the flow. Chaotic systems are also characterized by intermittency, or cyclic behavior through a very long series of states. A graphic illustration comes from Mandelbrot's fractals (Figure 1.8). His "San Marco Dragon" is chaotic because the Dragon traces a "long orbit of successive positions" (1982: 198), points plotted according to a formula that, as successive values are plugged into it to yield coordinates, but eventually returns to the origin and forms the graphic design. The scaling behavior of the Koch island (Figure 1.3), on the other hand, or the scale-free network of villages, towns, and cities in the economic market, is well ordered and much simpler by comparison, just aggregation of more and more of the basic patterns, forming new patterns at higher levels of scale.

Descriptions of self-organization typically use data collected in time series, and apply complex mathematical operations to generate "attractors," as shown by Guastello and Liebovitch for psychology (Figure 1.9). As for the application of successive values to make Mandelbrot's San Marco Dragon with a formula, successive measurements of real phenomena over time may create patterns when graphed that tend towards a fixed point (A), or create an oscillation or orbital shape (C, E), or other patterns whose regularity may be more difficult

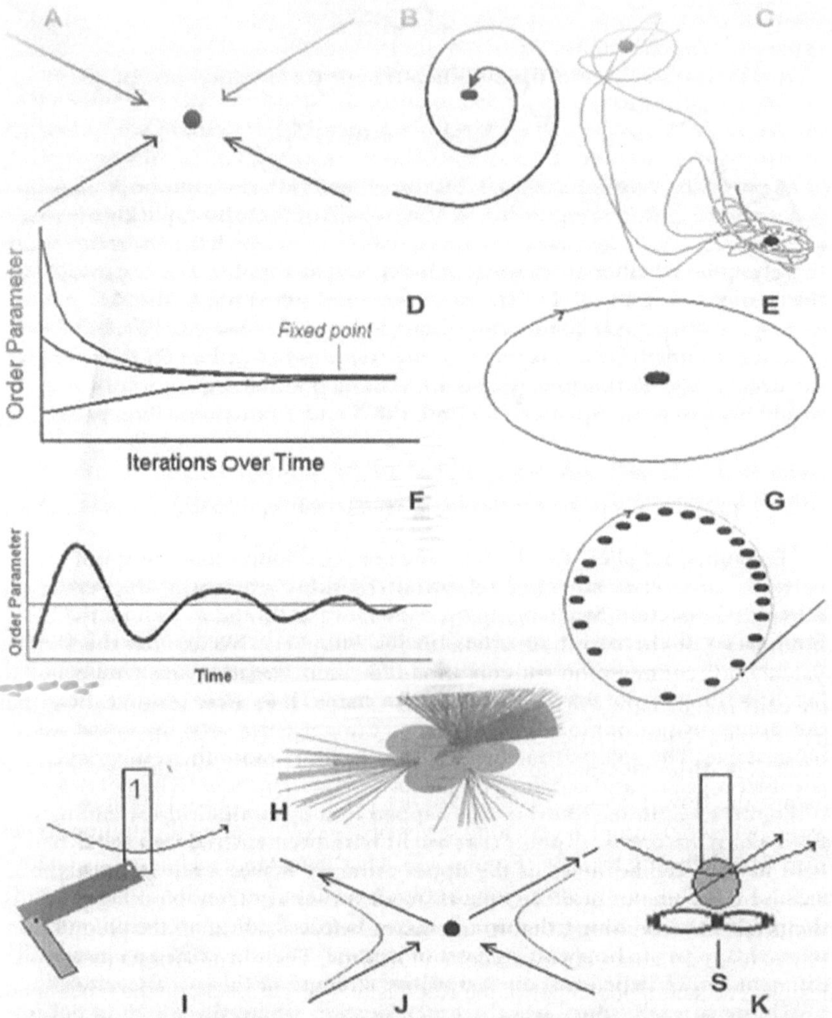

Figure 1.1. Gallery of basic dynamics: A, fixed point attractor (Eqs. 1.1 and 1.2); B, spiral attractor (Eqs. 1.3); C, trajectory of a control point in a field containing two fixed points; D, time series of points gravitating to a fixed point; E, limit cycle (Eqs. 1.4 and 1.5); F, time series for a dampened oscillator (Eq. 1.6); G, limit cycle created by organizing fixed points into a circle; H, system of three repellors; I, three coupled oscillators; J, saddle point; K, saddle created by a perturbed pendulum.

Figure 1.9 Basic dynamics (Guastello and Liebovitch 2009)

to see (chaotic, or "strange" attractors). Each successive moment in time corresponds to a "state" of the phenomenon being observed, whether it is traffic in a city or economic activity in a country or evolutionary development in a biological system. Stuart Kauffman (1995: 80–92) has shown with simulations of successive states, in his case patterns of lighted bulbs in a network of 1,000 light bulbs, that complex systems with self-organizational order settle into a very small number of states that are insensitive to small changes, while chaotic systems occupy a very large number of states – a number literally millions of times more than those in the complex system. So, in some cases it may take a very short time to reach an attractor (which is really just another term for regularity or ordered behavior achieved by the elements being observed), but in other cases it may take a very long time indeed. Stephen Wolfram (2002) investigated all of the possible rules that can be created for a one-dimensional cellular automaton (like the Game of Life, only with just eight cells all in one row), and he found that most rule sets quickly create a fixed state; some rule sets create oscillating patterns ("blinkers"); a smaller number of rule sets creates chaotic behavior; and the smallest number of rule sets creates complex behavior, the order that emerges from a complex system. The self-organization that we see among the ants, in the Game of Life, in whirlpools and rivers, in economic markets, and in many other circumstances all around us thus arises only rarely among all of the things that could happen in a large network of interacting components, but emerge it does. In Stuart Kauffman's colorful phrase (1995: 80–81), the order that emerges from complex systems and conditions so many aspects of our lives arises "at the edge of chaos."

So, we can say again that complex systems show (1) continuing dynamic activity in the system, (2) random interaction of large numbers of components, (3) exchange of information with feedback, (4) reinforcement of behaviors, (5) emergence of stable patterns without central control. The activity in complex systems is open and dynamic, as opposed to a static structure. The many components in a complex system interact randomly, as opposed to standing still in a hierarchical arrangement of types. While complex systems depend upon the exchange of information, probabilistic feedback keeps the system from becoming stuck in rule-bound relations. Reinforcement of behaviors from feedback creates nonlinear distribution of units, as opposed to random or statistically normal distribution, at every level of scale instead of at just one level characterized by homogeneous unity.

Language as a complex system

We have never used time series data very much for language studies, because unlike the binary on/off data from many sciences, our data is "rich," containing lots of information per data point, and so it is expensive to collect and analyze:

it is impractical to go back and repeat observations frequently enough for time series. We do look back in time in historical linguistics to observe change, but we do not have enough information about how the members of a population were actually speaking at any given time to make any more than speculative judgments about the state of the language in the remote past (see Chapters 3–5 for discussion of "grammaticalization," and for discussion about how a knowledge of complex systems affects our historical view). The situation is little better for the recent past, when we may have more information from writing or even from recorded speech, but still no fair way to estimate how all of the members of a population were actually speaking (see Chapter 7).[4] This means that we need to look for the effects of complex systems in speech in the surveys and other collections of data that we can actually carry out.

For language, as for other domains in which complex systems operate, complex systems allow the stable patterns to emerge on which we rely for effective communication in our social interactions. The long state cycles of Mandelbrot's Dragon do not account for our universal impression of different kinds of speech around us, so chaotic systems are not something we should expect to describe language in use; such descriptions would be little better than the old idea of free variation (see Chambers 2009: 12–14). Complex systems, on the other hand, are well suited for descriptions of language varieties. It seems clear, however, that linguistic rule structures as usually described in grammars, or in Bybee's 2001 phonology, do not meet these criteria. By their nature, rule systems are static, hierarchical structures at low-energy equilibrium, which is the whole point of generalizing away from the complex, energetic interactions of language in use (as Saussure and others have argued; see Kretzschmar 2009: ch. 2). However, if we pay attention to the interactions themselves of language in use, we will see that speech does self-organize into ordered but dynamically changing states as a complex system.

Let us begin with what we know. The fact that speech is open and dynamic has never been a matter of dispute. As far back as Wenker and Saussure, we have never been able to find natural boundaries for dialects, and Saussure says the same even for languages themselves (these matters are also treated in detail in Kretzschmar 2009: ch. 2). New conversations and new writings among members of any speaking population occur continuously at unimaginable rates,

[4] The problem of rich data also introduces greater dimensionality for the description of systems. Stephen Wolfram (2002) required over 1,000 pages to prepare a comprehensive description of the patterns created by the successive states of a one-dimensional cellular automaton (a set of eight boxes in a row, which can be either black or white), according to the application of different rules for how the on/off pattern would change between states. Kauffman's light bulbs were arranged in a two-dimensional grid. Speakers as agents interact with each other in many different ways, and speakers can choose between many possibilities for any linguistic feature we wish to observe.

surely a dynamic system. If these things no longer occur for a language, we call it "dead," like Latin or Old English (when consulted recently our Web friend Wikipedia offered a list of nearly 100 "recently extinct languages," *s.v.* 'extinct language'). This term parallels what Stuart Kauffman has said of biological organisms, that for them such a low-energy equilibrium state "corresponds to death" (1995: 21). Moreover, we do see self-organization from speech as we would expect from a complex system, in the form of geographical, social, and textual clustering of speech sounds and words. We recognize that sounds and words are not randomly distributed in speech interactions, but instead are associated in different ways with particular localities, particular social groups, and particular text types. Speech clearly meets these initial conditions for complex systems.

Let us turn now to some experimental evidence, which I draw from my American Linguistic Atlas Project (for information, see Kretzschmar et al. 1993, and www.lap.uga.edu), beginning with the large number of interactive components required for complex systems. If we take the interactive components of speech to be words and speech sounds (leaving clauses aside for the moment since my data has less evidence for those; see the discussion of "constructions" in Chapter 3), there are actually a great many more components than we generally recognize in the traditional formulation of linguistic systems. The number of entries for headwords in the dictionary, for example, grossly underestimates the possible units in the lexicon if we take multi-word collocations into account. In our LAMSAS survey of 1,162 people, we typically find large numbers of lexical variants, many of them multi-word units. For instance, in Table 1.1 for *dry spell* 'weeks without rain' we can still count 39 different variants even after we get rid of possibly inappropriate responses (like *blessing* or *month*) and incidental differences between responses. There are a great many more than what we might have expected, *drought* or *dry spell*. Many of these responses share adjectives like *clear* or *dry*, and many more share nouns like *spell* or *weather*, and we might be tempted to try to reduce the variation by grouping responses according to shared elements. However, each entry as a combined form is a separate type of response, and in the study of complex systems we are interested to retain the variation, the full range of interactive components, rather than trying to reduce it. In Table 1.2 for *parlor* 'place to meet guests' we count 99 different variants. As Allison Burkette has shown (2012), many of these variants have arisen from different modes of American house construction, and some terms, like *morgue*, arose from cultural practices such as laying out the dead in the front room of the house. The variation in terms does not come just from slips of the tongue, although of course those happen, but instead encodes real variation in cultural patterns and practices. While it is true that not all of the different words may refer to exactly the same thing, it is also true that no two uses of the same word may refer to exactly the same thing. In a survey like

Table 1.1 Dry spell *adjusted for plurals, inappropriate responses (39)*

a long dry spell	1
a spell of weather	1
bad drouth	1
bad dry spell	1
clear spell	1
clear spell of weather	1
clear weather	1
draft	1
drought	164
drouth (s)	719
drouth spell	1
drouthy spell	1
dry	5
dry and dusty spell	1
dry drought	14
dry drouth	15
dry hot spell	1
dry season	10
dry spell(s)	905
dry spell like	1
dry time	25
dry weather	38
dry weather spell	1
fair weather	1
famine	2
going to get dry enough to	1
It's some kind of dry	1
long drouth	2
long dry spell	6
one little dry spell	1
parching the corn	1
real drought	1
real drouth	2
real dry spell	1
spell	1
spell of dry weather	8
two weeks of clear weather	1
very dry spell	1
warm spell	1

the Atlas, we rely on the fieldworkers to give us answers that match the prompt, whether that is 'weeks without rain' or 'place to meet guests,' and then accept the reality and comparability of all of the different responses they record as components in the complex system of speech. We also get large numbers of unique types in our Atlas IPA transcriptions of words, 242 different realizations

Table 1.2 *LAMSAS Tally for* parlor *'place to meet guests' adjusted for plurals, inappropriate responses (99) 12 back parlor*

12	back parlor	1	middle room
2	back room	1	morgue
3	bedroom	3	music room
1	best parlor	1	no man's lands
4	best room	1	office
46	big house	1	organ
1	big living room	1	other room
17	big room(s)	629	parlor(s)
1	buckra setting hall	1	parlor cook stove
8	chamber	1	parlor floor
2	chamber room	1	parlor matches
1	colonnade	12	parlor room
15	company room	1	parlor stove
1	cutting room	1	parsonage
1	dance	1	preacher's room
10	den(s)	1	quarter
1	den place	5	reception room
6	dining room	1	red room
2	double parlor	62	rented (me) a room
30	drawing room	10	room(s)
1	east hall	1	rumpus room
1	eating room	1	second drawing room
1	entrance hall	294	setting room
1	entry	1	sewing room
2	family living room	1	shed room
4	family room(s)	461	sitting room(s)
1	family sitting room	1	social room
7	fire room	1	solarium
1	first story	1	spoorruim
1	foyer	1	study room
12	front parlor	1	summer parlor
161	front room	1	sun parlor
4	great house	1	tea room
2	guest room	1	television room
11	hall	1	the big house
1	hall room	1	the big room
7	hired (me) a room	1	the front room
1	home room	4	the hall
3	house	5	the house
1	I ain't going to say 'living room'	1	the house part
1	in the hollow of the house	17	the room
1	in the house	1	the upper room
1	large room	1	two parlors
1	levenruim	1	visiting room
19	library	1	waiting room
1	little sitting room	1	warm room
625	living room(s)	1	withdrawing room
1	lounge room	1	work room
1	main house	2	yellow room
1	middle hall		

of the vowel in *fog*, when we consider things like diphthongization and length in addition to degrees of height and frontness (Table 1.3). Again, we have to rely on the field workers faithfully to record what they heard – and field workers like Guy Lowman and Raven McDavid were highly trained and experienced – even if we wonder whether most people could distinguish between all of the different pronunciations in practice. The field workers used all of the facilities of the IPA; they did not just put down what they may have expected to hear. Even for the more stable vowel in *six*, Atlas field workers recorded 29 unique types of vowel realization (Table 1.4). The vowel in *six* is often thought to be invariant, but that was clearly not the case for the Atlas field workers. Acoustic phonetics, of course, the use of computers to plot speech sounds according to their principal formants (F1, F2, sometimes others), replaces the small number of IPA vowels with infinite possibilities for variation because of its continuous measurement of frequency for each of several formants.

The 20 base IPA symbols in our Atlas set plus diacritical marks can yield only a small number of variants in comparison. We could, of course, try to reduce this variation by saying that the pronunciation of *fog* is different from the pronunciation of *fig*, *fag*, and *fugue*, or that the pronunciation of *six* is different from the pronunciation of *seeks*, *sex*, *sax*, *socks*, *sucks*, and *soaks*. It turns out to be more difficult than we might expect to sort variants into exclusive groups for each of these alternatives (see Kretzschmar 2012 for discussion of the notion of contrast). As for the variants of *dry spell*, however, when we think about complex systems we are interested in preserving the full range of interactive components. So, experimental evidence again shows us that speech has the large number of interacting components that we expect of a complex system, if only we do not artificially constrain how we classify its variation.

Perhaps the most striking evidence for speech as a complex system is the nonlinear distribution of the variants for any given linguistic feature. Linguists will recognize Zipf's Law, a frequency ranking of words in texts that always finds that rank is roughly inversely proportional to frequency. That is, there are a few very frequent words and some moderately frequent words, but most words occur at very low frequencies. This yields the curve at left in Figure 1.10 for the original ranking, and the straight line at right if the logarithms of the ranked frequencies are plotted. It turns out that this nonlinear distribution, which I will call the A-curve for short, is not just a property of texts, but occurs literally everywhere in our Atlas survey research on language in use. So, the A-curve occurs for the 39 lexical variants for *dry spell* (Figure 1.11). It also occurs for the 99 lexical variants for *parlor* (Figure 1.12). It also occurs for the 242 phonetic variants for the vowel in *fog* (Figure 1.13). And it occurs for the 29 phonetic variants for the vowel in *six* (Figure 1.14). These curves are all a little different from each other (aspects of measurement of the curve are discussed in Chapter 7), but they are all alike in the basic property of having just a few

Table 1.3 *Variation in the* fog *vowel ({} indicates weak realization)*

fɑ·g	93	fɒ∧og	5
fɒ∧·g	72	fɔ∨{ə}g	5
fɔ∨·g	71	fɔ<·{ə}g	5
fɒ̜∧<{ə}g	46	fɔ∧<·{ə}g	5
fɑ∧>{ə}g	44	fɔ∨<·{ə}g	5
fɒ∧<{ə}g	44	fɒ̜<·{ə}g	4
fɔ·g	43	fɒ̜∧·g	4
fɑg	27	fɒ̜∧·g	4
fɔ∨ɔg	27	fɒ<g	4
fɑ<·g	26	fɒ<g	4
fɒ·g	22	fɒ̜∧<g	4
fɔ:g	21	fɒ∧<{ɒ∧}g	4
fɑ·{ə}g	19	fɒ∧<{ɒ∧}g	3
fɒ∧<·{ə}g	18	fɒ∧{ɔ∨}g	3
fɒ∧·{ə}g	18	fɑ{ə}g	3
fɑ∧>·{ə}g	18	fɒ̜̥{ə}g	3
fɑ∧{ə}g	17	fɒ̜̥{ə}g	3
fɒ∧ɔ∨g	16	fɑ{ə}g	3
fɒ<·g	15	fɒ̜·{ə}g	3
fɔ·{ə}g	13	fɒ·ɔ∨g	3
fɒ̜·{ə}g	12	fɒ·ɔ∨g	3
fɑ>·{ə}g	12	fɒ̜g	3
fɒ∧·ɔg	12	fɔ∧{ə}g	3
fɑ∧·g	12	fɔ·ɔ∧g	3
fɔɔ∧g	12	fɔ∨g	3
fɒ̜{ə}g	11	fa>:g	2
fɑ∧>g	11	fɒ∧{ɔ}g	2
fɒ∧{ə}g	10	fɑ∧<{ə}g	2
fɒ∧·{ɔ∨}g	10	fɒ̜∧>{ə}g	2
fɔ∨·{ɔ}g	10	fɒ̜∧>{ə}g	2
fɔ∨·{ə}g	10	fɒ·{ɔ}g	2
fɑ∧g	9	fɒ∧·{ɔ}g	2
fɔ∨{ɔ}g	9	fa·{ə}g	2
fɒ̜∧<·{ə}g	8	fɒ̜̥·{ə}g	2
fɔ·{ɔ∧}g	8	fɒ̜∧·{ə}g	2
fɒ∧:g	7	fɒ·{ə}g	2
fɒ̜<{ə}g	7	fɒ̜∧<·{ə}g	2
fɑ>{ə}g	7	fɒ̜<·{ə}g	2
fɔ∨:g	6	fɒ·ɒ∧g	2
fɒ{ə}g	5	fɒ̜·g	2
fɒ̜∧{ə}g	5	fɒ̜·g	2
fɑ∧·{ə}g	5	fɒɒ∧g	2
fɒ̜<· g	5	fɒ∧<ɔ∨g	2
fɒ∧<·g	5	fɑ̆∧>g	2
fɒ∧<·g	5	fɑ>g̥	2
fɒ∧ɔg	5	fɑ∧<g	2
fɒg	5	fɔ∧:g	2

(*cont.*)

Table 1.3 (*cont.*)

fɔ{ɔ∧}g	2	fɒ̯{ʌ∧>}g	1
fɔ∨{ɔ̯}g	2	fɒ·{ɒ∧}g	1
fɔ<{ə}g	2	fɒ∧<·{ɒ∧}g	1
fɔ∨·{ɔ∧}g	2	fɒ∧·{ɔ<}g	1
fɔ ·{ə̯}g	2	fɒ̯∧·{ɔ∨}g	1
fɔ∧·{ə}g	2	fɒ̠<·{ɔ}g	1
fɔ∧·{ə̯}g	2	fɒ̠ ·{ə}g	1
fɔ̯∧·{ə}g	2	fɒ̯∧·{ə̯}g	1
fɔ̯∧·{ə̯}g	2	fɒ̠∧·{ə}g	1
fɔ∨<ɔg	2	fɑ·{ə∧>}g	1
fɔog	2	fɒ̯·{ə̯}g	1
fɔ∨og	2	fɒ̯·{ə>}g	1
fo{ʊ}g	2	fɑ>·{ə̯}g	1
fɒ∧:{ə}g	1	fɒ̯∧<·{ə}g	1
fɒ̯∧:{ɣ}g	1	fɒ̯∧<·{ə̯}g	1
fɒ̯:g	1	fɒ∧·{ə̯}g	1
fɒ:og	1	fɒ̮ ·{ə}g	1
fɒ{ɒ̮ ∧<}g	1	fɒ·ɔg	1
fɒ{ɒ̯∧}g	1	fɒ̠·ɔg	1
fɒ<{ɒ∧}g	1	fɒ∧·ɔ∨g	1
fɒ̮ {ɒ}g	1	fɒ̯·əg	1
fɒ̮ {ɒ∧}g	1	fɒ̠ ·g	1
fɒ̯<{ɒ̮ ∧}g	1	fɑ̄·<g	1
fɒ̯<{ɒ̯∧<}g	1	fɒ̯∧<·g	1
fɑ>{ɒ̮}g	1	fɑ∧<·g	1
fɒ{ɔ∨}g	1	fɒ̠·og	1
fɒ̯∧<{ɔ∧<}g	1	fɒ∧·o∨g	1
fɒ̯{ɔ}g	1	fɒ̠<·o∨<g	1
fɒ<{ɔ∨}g	1	fɒ̠∧·o∧g	1
fɒ∧{ɔ}g̥	1	fɒ∧<·og	1
fɒ∧{ɔ̯∨}g	1	fɒ̠ɒg	1
fɒ̮ {ɔ}g	1	fɒ̯<ɒ∧<g	1
fɒ̯∧{ɔ̯∨}g	1	fɑ>ɒ∧<g	1
fɒ̯∧{ɔ̠∨}g	1	fɒɔg	1
fɒ∧<{ɔ∧}g	1	fɒɔ∨g	1
fɒ∧<{ɔ∨}g	1	fɒ̠ɔg	1
fɒ{ə∧>}g	1	fɒ̯<ɔ∨g	1
fɒ̯∧<{ə}g	1	fɒ̠∧ɔg	1
fɒ̯{ə}g	1	fɒ̠∧ɔ∨g	1
fɒ̯∧{ə}g	1	fɒ̯<g̥	1
fɒ̯∧<{ə∧>}g	1	fɑ̆g	1
fɑ>{ə̯}g	1	fɒ̦g	1
fɒ∧{ə}g̥	1	fɑg̥	1
fɒ∨{ə}g̥	1	fɑ<g	1
fɒ̯<{ə∧>}g	1	fɒ∧g	1
fɑ∧{ə>}g	1	fɒ̮ g	1
fɑ∧>{ə>}g	1	fog	1
fɒ∧<{o∨}g	1	fɒ<o∧<g	1

Table 1.3 (*cont.*)

fɒ∧o∨g	1	fɔ∨·{o∨}g	1
fɒ̯∧og	1	fɔ∧·{ʊ}g	1
fɒ∧<og	1	fɔ∨·ɔ∧g	1
fɔ∧:{ə}g	1	fɔ∨<·ɔ<g	1
fɔ∨:{ə}g	1	fɔ·əg	1
fɔ̯:g	1	fɔ̯∧·əg	1
fɔ̯∨:g	1	fɔ·g̥	1
fɔ{ɔ̯}g	1	fɔ̯·g	1
fɔ{ɔ̯∧}g	1	fɔ̯∧·g	1
fɔ∧{ɔ̯∧}g	1	fɔ∧<·g	1
fɔ∨{ɔ∧}g	1	fɔ·og	1
fɔ∨{ɔ̯∨}g	1	fɔ∨·og	1
fɔ∨<{ɔ<}g	1	fɔɔ̯g	1
fɔ{ə}g	1	fɔ̯∨<ɔg	1
fɔ̯∧{ə̯}g	1	fɔ∨ɔ̯g	1
fɔ∧{o∨}g	1	fɔ∨ɔ∧g	1
fɔ∨{o̯}g	1	fɔg	1
fɔ∨{o∨}g	1	fɔ<g	1
fɔ·{ɔ}g	1	fɔ∧o{g}	1
fɔ·{ɔ̯}g	1	fɔ∧og	1
fɔ·{ɔ̯∧}g	1	fɔ∨o∧g	1
fɔ̯·{ɔ̯∧}g	1	fo∨{ʊ∨}g	1
fɔ̯·{ə}g	1	fo·g	1
fɔ̯·{ə̯}g	1	fog	1
fɔ̯∨·{ə̯}g	1	fo̯∨<ʊg	1

Table 1.4 *Variation in the* six *vowel ({} indicates weak realization)*

sɪks	604	si∨{ᵻ}ks	1
sɪ∧ks	213	sɪ·{ə>}ks	1
sɪ·ks	129	sɪ∧·{ə}ks	1
sɪ∧{ə}ks	69	sɪ·{ᵻ}ks	1
sɪ>ks	24	sɪ∨·ks	1
sᵻks	17	sɪ∧>·ks	1
sɪ∧·ks	16	sɪ<ɪks	1
si∨ks	10	sɪ<ks	1
sɪ{ə}ks	11	sᵻ∧ks	1
sɪ∧>ks	9	sᵻ∨ks	1
sɪ∨>ks	5	sɪ∧<ks	1
sɪ∨{ə}ks	2	sɪ∧>·ks	1
sɪ∧{ᵻ}ks	4	sɪ<ɪks	1
sᵻ<ks	4	sɪ<ks	1
sɪ·{ə}ks	3	sᵻ∧ks	1
sĭks	2	sᵻ∨ks	1
sɪ{ə>}ks	1	sɪ∧<ks	1
sɪ∧>{ə}ks	1		

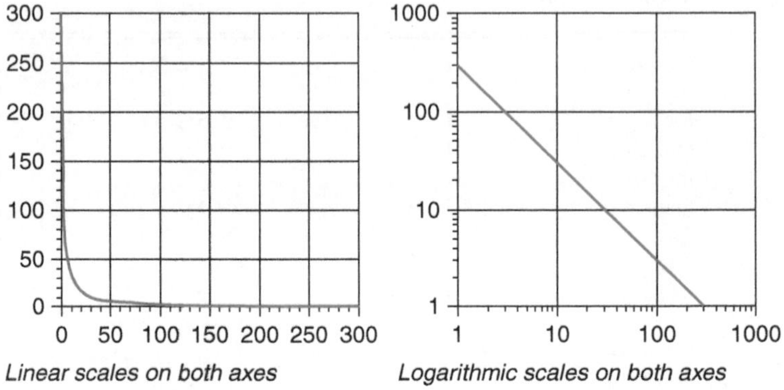

Figure 1.10 The distributional pattern: "Zipf's Law," "power law," asymptotic hyperbolic curve of ranked frequencies, aka "A-curve" (cited from www.nngroup.com/articles/zipf-curves-and-website-popularity, an article about website popularity)

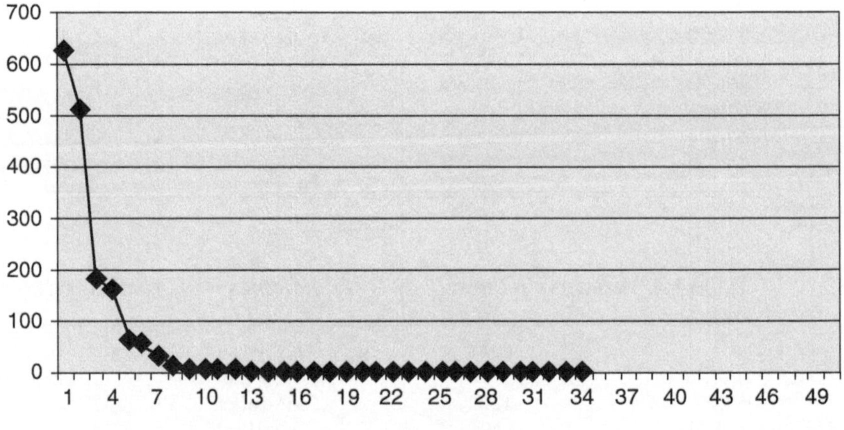

Figure 1.11 A-curve, *dry spell* variants

common variants, some moderately common ones, and many uncommon ones. Indeed, the A-curve distribution occurs not just in every item in our LAMSAS survey, it occurs in every item in every survey I know of. My students who have replicated parts of American Atlas surveys, like Ellen Johnson (1996), Allison Burkette (2001, 2009), and Susan Hoover (2001), have found the A-curve throughout their data. Surveys in new American regions show it (Hamilton-Brehm 2003 in El Paso, Antieau 2006 in Colorado). Hans Goebl has told me that a distribution of this kind is what he expects from his work on English, French, and Italian surveys (p.c.). A student of mine who studies Polish variation finds

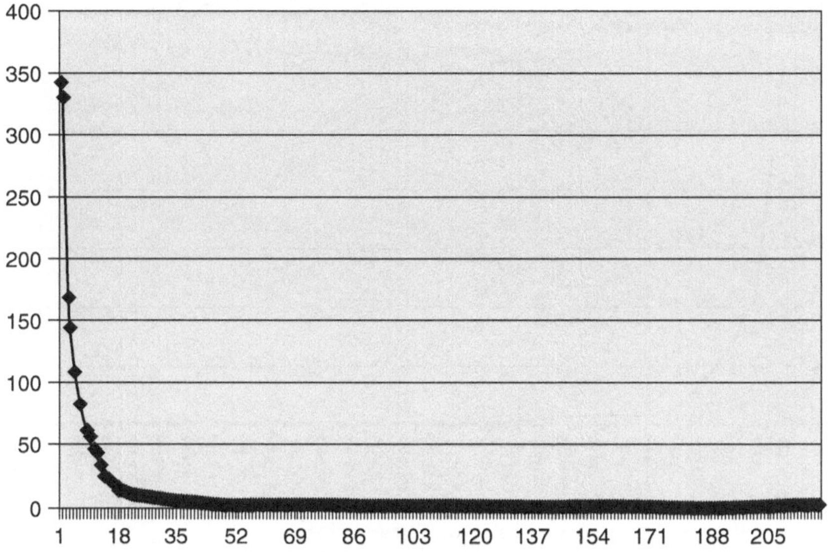

Figure 1.12 A-curve, *parlor* variants

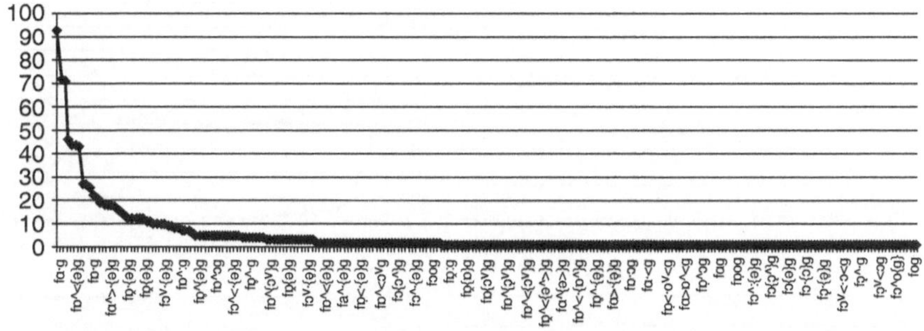

Figure 1.13 A-curve for *fog* vowel realizations

the A-curve distribution routinely in her work (Bounds 2010), and a student who did a survey of the lexicon of Northern Thailand has found the A-curve in her Thai variants (Hatfield 2005). Mello (2013) found it throughout a corpus analysis of the syllables of Vietnamese. If we put our survey research findings together with Zipf's Law from textual data, the nonlinear distributional pattern of complex systems does certainly appear to belong to speech.

As for the scaling property of complex systems, we regularly observe that, too. So, for *bureau* 'case furniture in which to keep your socks' (Figure 1.15; see Burkette 2001, 2009) we see the A-curve for LAMSAS overall (upper left),

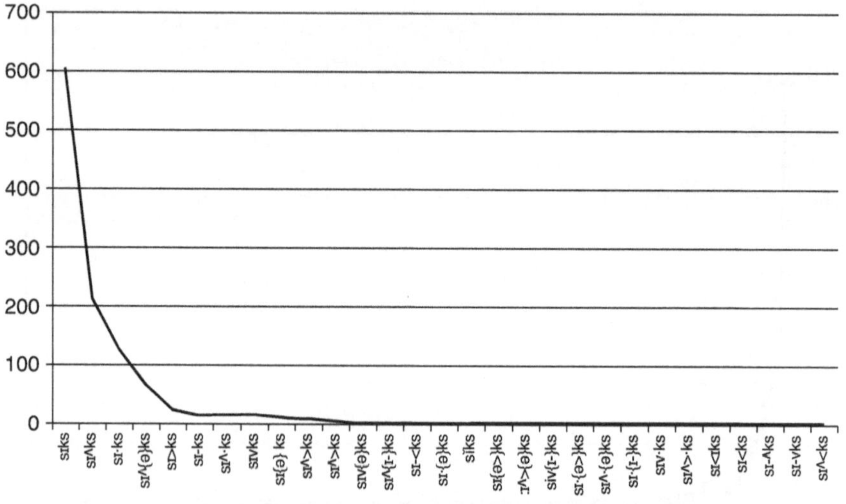

Figure 1.14 A-curve for *six* vowel realizations

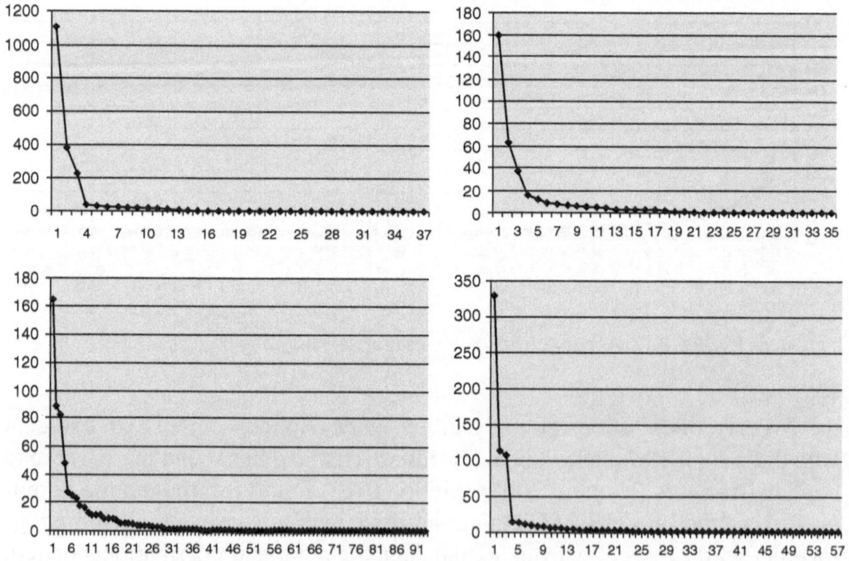

Figure 1.15 Scaling property of A-curve: *bureau* variants for all, SC/GA/FL, NY, women

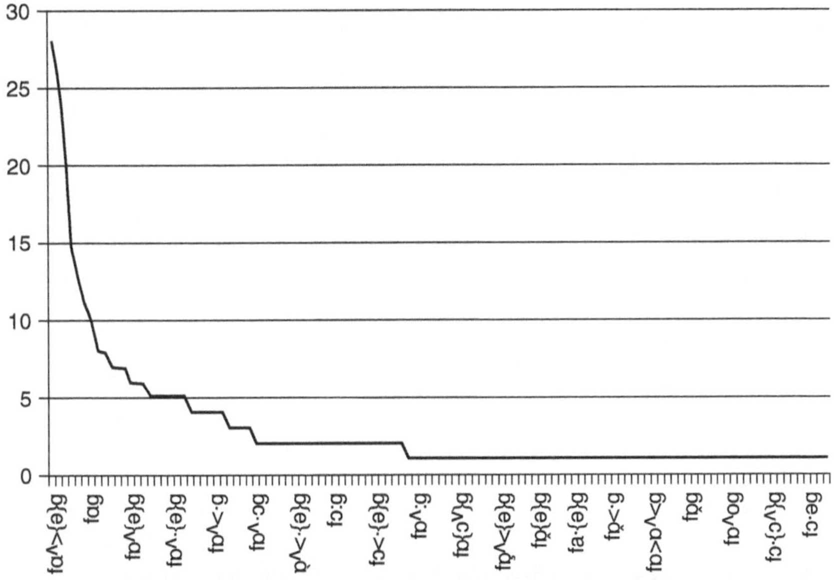

Figure 1.16 Scaling property of A-curve: *fog* variants for women

and then in subsamples for the Southern region (SC/GA/FL), for New York State, and for a social group, the women speakers of LAMSAS. Instead of regional subsamples, we can choose to observe socially defined subsamples and see the same effect. Figure 1.16 is a view of the women's subsample of vowel realizations for *fog*, which shows not only that the A-curve works with phonetic subsampling but also that the particular realizations even at the top of the list are not the same for the whole survey and for the subsample. Exactly the same thing happens for the subsample of speakers just from Type II, essentially all those speakers with a high-school education (Figure 1.17). All of these illustrations are clearly recognizable as a curve with a few common variants, some in the middle, and many infrequent variants in the long tail of the curve.

We now have an answer for Edgar Schneider's question about chaos theory and speech. No, chaos theory is not a model for dialect variability and change. But speech does constitute a complex system, "at the edge of chaos." Observations over time are still of interest, but not required as Chambers had speculated. Our existing data shows us the nonlinear A-curve distributional pattern that, along with the scaling property of speech, provides experimental confirmation to accompany the connections of language variation with complex systems that were already apparent to us.

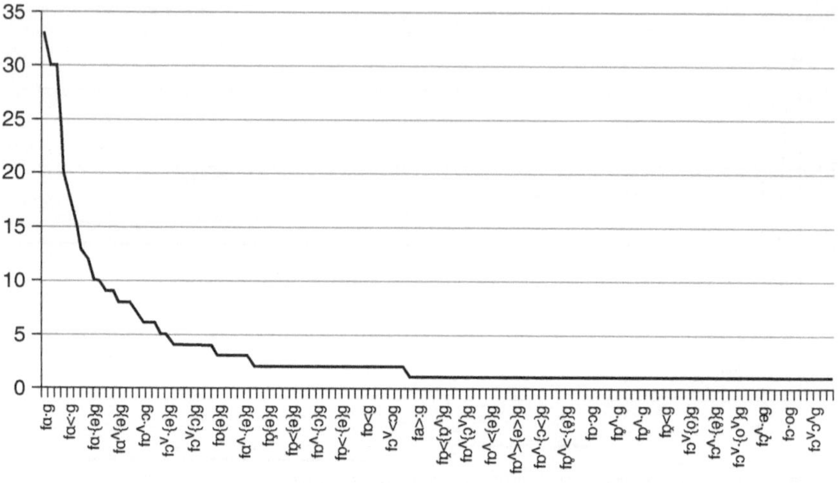

Figure 1.17 Scaling property of A-curve: *fog* variants for Type II speakers

We can use this stable underlying A-curve distribution to address what has been one of our biggest problems: defining the relationship between what people actually say or write and the generalizations that we want to make from that behavior. The key aspect of the A-curve is that it gives us, the users of speech, a perceptual aid so that we can make sense of the speech interactions around us (Figure 1.18). Speakers perceive the most common variants on the A-curve for each linguistic feature as "normal" or "expected," and the long tail of infrequent variants as "different." What matters is the sharp difference in frequency between variants, not the precise statistical fit of the curve to a formula (see Chapter 7). As we have seen, the A-curve is always present for any population we sample or subsample, so the distribution itself gives users of speech consistent, high-quality input for such perceptions (Ellis and Larsen-Freeman 2009a suggests that this distribution helps us in language acquisition). So, for example, a diphthongal pronunciation of *fog* is what we expect from women, but not what we expect from the LAMSAS speakers overall. We can expect 'weeks without rain' to be called a *dry spell* as a normal word, but accept that many other words are possible, and understandable, as different variants. We can perceive the top-ranked variants of any linguistic feature for groups at any level of scale, and the fact that different variants for a given feature will be ranked more highly in different groups helps us to distinguish the language behavior of the group. At the same time, speakers are not bound to use only the top-ranked variant for a group since many other variants will be in use. Human agents, unlike ants, can think about and choose how to deploy linguistic variants, but that does not change the basic operation of feedback

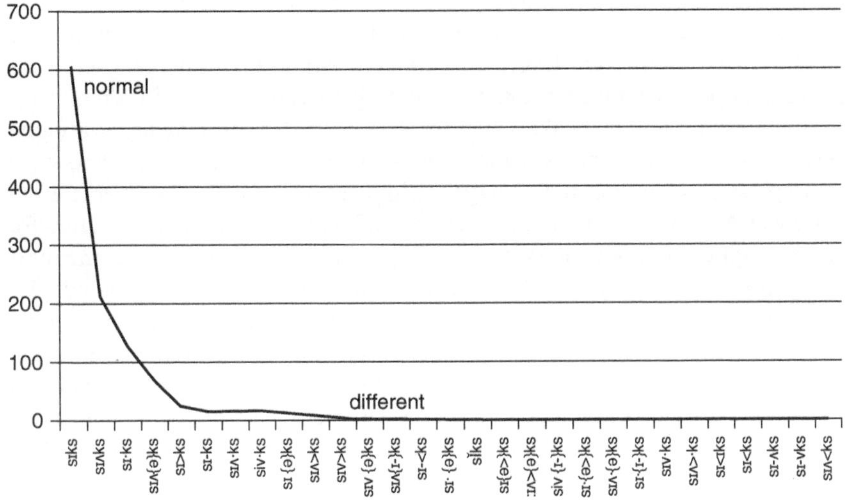

Figure 1.18 Perceptual aid from the A-curve

and reinforcement; we make choices in relation to the order arising from the complex system. Moreover, speakers will belong to many different groups all at the same time (regional, social, and interest groups of many kinds in scale-free networks), and so speakers will always be in the position of choosing the variant for any feature, normal or different, that they judge to be best for any given linguistic event and set of interlocutors.[5] We humans are smarter than ants, and we can command a great many more behaviors in language as they emerge in all of the different regional, social, and textual settings for speech.

The perception of aggregated collections of "normal" variants for many features at some level of scale, whether it is a geographical region, a social or interest group, or a text type, creates what we can call "observational artifacts" (following Günther et al. 1996, a group of physicists). That is, linguistic systems, whether for large populations of speakers or for smaller groups of speakers, are objects whose existence comes from our perception of reality, not from reality itself. For any group of speakers and for any linguistic event, we can mentally assemble the top-ranked variants for many features into what we call a linguistic system, at the same time that we know about and understand

[5] It goes without saying, at least in this book, that speakers make their choices at a particular moment in time and that, while the A-curve distributional pattern is always present, particular variants become more or less frequent over time – in a manner of speaking, they climb up or fall down the curve. Kretzschmar and Tamasi (2003) offers a detailed account of this (though without reference to complex systems), and Kretzschmar (2009) adds additional discussion. Chapter 5 in this volume addresses the point.

many variants of lesser frequency of use among the speakers in any group and for any event. As we have also seen, different variants of a feature may be top ranked at different levels of scale even within the same data set. Thus the scaling property in the complex system of speech accounts for the universal impression among speakers that there is characteristic language variation between different regional or social groups, at the same time that speakers have the impression of an overall language at a higher level of scale. The ever-present nonlinear A-curve distributional pattern actually provides good justification for our description of linguistic systems, as a generalization that captures the most frequent usages for any and all linguistic features by speakers associated with a group, so long as we understand that our generalization can only be accurate for that specific group of speakers. We can construct an accurate grammar for any group at any level of scale – but the problem of scale dependency makes it certain that a single usage-based grammar will never fit all groups at all levels of scale. Varieties of natural language have no contract that speakers all agree to; rather, each of us perceives normal variants for features of the language behavior of different groups of people that are important to us, and we make our linguistic choices based on those perceptions.[6] Each of us makes our own observational artifacts from the language that we experience.

When we as users of speech perceive a frequent variant for a group of speakers, we commonly make the generalization that that variant is "normal" for everybody in that group and so everybody must habitually use it. We may think that American Southerners, for instance, must all say *y'all* for *you* plural, and say *far* and *tar* for *fire* and *tire*, as part of their linguistic system. Such thinking is always an overgeneralization. The fact that we can always identify top-ranked variants for any feature does not mean that top-ranked variants are categorical. Indeed, the distribution of variants in the group in reality always has an A-curve pattern, and so the application of a categorical rule always misrepresents the facts of language in use. *Y'all* is a very frequent variant for *you* plural in the South but not everybody uses it, and there is actually lots of variation in how people pronounce *far* and *tar* for *fire* and *tire* there.

What those interested in language now should get used to, from the evidence of complex systems, is that any variety we name actually exists as an observational artifact that comes from our perceptions of the available variants. Frequency effects do not generate linguistic systems in reality, as linguists have described them for the last century. Linguistic systems as low-energy equilibrium systems are always observational artifacts of our perception, in effect

6 "Standard" language is not the same as natural language, but instead is an institutional construct (see Kretzschmar 2009: 223, 262–263, 271–277). It is entirely appropriate that sociolinguists measure the linguistic systems that they report for social groups, but comparison with a "Standard" is not as simple as it is sometimes thought. See Chapter 7.

transformations of speech data from its natural existence as part of a complex system. When Östen Dahl discusses "linguistic complexity" (2004), he is interested in how detailed and involved our description of a formal, hierarchical system of language might be, clearly a different sense of complexity from our discussion of complex systems. When we propose the existence of such hierarchical linguistic systems based upon our perception of speech around us, as we certainly want to do and are justified to do by the distributional patterns of speech as a complex system, we need to be guided by the nonlinear distributional pattern of the evidence of language in use because no system that we describe is actually instantiated in the spoken interactions themselves. There are no contracts for natural language. The finding that speech is a complex system gives us a better framework to understand it, a framework we can share with other sciences, and one that we can apply to the theoretical and practical problems of interest to us. The next chapter deals with scientific epistemology, how the recognition of complex systems as a model for language fits with trends in science beyond linguistics. The following chapters are designed to do just that on a range of topics, from usage-based linguistics, grammars, and history of the English language, to chapters on cognitive issues and sociolinguistics.

2 Linguistics, science, the humanities, and complex systems

As we have seen, empirical findings from linguistic survey research support the idea that speech is a complex system. The question remains, however, how we should incorporate this idea into our contemporary expectations for scientific study in a discipline. Nobody in academia today can doubt that that the physical and natural sciences, with applications in engineering and the health sciences, have now surpassed the humanities and many parts of the social sciences in their importance to our educational institutions. We can hear complaints about the social sciences as "soft sciences" or "fuzzy sciences," which attacks the legitimacy of such study when compared to the "hard sciences" that supposedly exhibit greater methodological rigor. The implication, perhaps, is that the social sciences and the humanities should adopt the methods of the physical and natural sciences in order to climb back up higher in the university hierarchy. The question takes on new dimensions when complex systems enter the picture.

Linguistics as a science

The Linguistic Society of America (LSA) maintains a webpage called "The Science of Linguistics" which offers a number of links to information about language study "from a scientific perspective," and also makes the following general statement (www.linguisticsociety.org/resource/science-linguistics):

> Modern linguists approach with a scientific perspective and methods what used to be thought of as a purely cultural academic discipline in the humanities. Contrary to previous belief, linguistics is multidisciplinary. It overlaps each of the human sciences including psychology, neurology, anthropology, and sociology. Linguists conduct formal studies of sound structure, grammar and meaning; investigate the history of language families; and research language acquisition. And as other scientists, they formulate hypotheses, catalog observations, and work to support explanatory theories.

From the point of view of understanding what *science* means, the passage requires some decoding. The *Oxford English Dictionary* defines the word (*s.v.*) under no fewer than five different main headings, each with subheadings.

The first sense, now current only in theology, refers to knowledge as a personal attribute, especially as opposed to moral conviction (*conscience*). The second sense is also individual, 'knowledge acquired by study' or 'trained skill' (as perhaps in boxing). The third sense, 'a recognized department of learning' especially as 'contradistinguished from art' (on this issue, see below), is more to the point for the LSA statement, which takes pains to separate modern linguistics from its earlier status as "a purely cultural academic discipline in the humanities." The LSA thus opposes "a scientific perspective and methods" to something "purely cultural" in outlook. It replaces its former status in the humanities by saying that the current discipline of linguistics overlaps with "the human sciences," a phrase of a type noticed in the *OED* as a means of classification for different disciplines by applying an adjective to the word *sciences*. The implication here is that, in each of these human sciences unlike the cultural disciplines of the humanities, the fourth definition in the *OED* has been achieved: 'A branch of study which is concerned either with a connected body of demonstrated truths or with observed facts systematically classified and more or less colligated by being brought under general laws, and which includes trustworthy methods for the discovery of new truth within its own domain.' When linguists "conduct formal studies of sound structure, grammar and meaning," they are enacting those trustworthy methods that can advance knowledge within the field. Linguistics is multidisciplinary according to the LSA, but it has its own identity on these terms. Finally, the LSA insists that the activities of linguists are those practiced by other scientists, including the formulation of hypotheses, collection and analysis of observations, and justification of explanatory theories. This statement also looks back in the direction of the *OED*'s fourth definition. The LSA thus wants to convey that linguistics has emerged from a questionable past in the humanities in order to join the general movement of modern science. The LSA does not address the fifth definition of the *OED*, that *science* is 'In modern use, often treated as synonymous with 'Natural and Physical Science,' and thus restricted to those branches of study that relate to the phenomena of the material universe and their laws, sometimes with implied exclusion of pure mathematics. This is now the dominant sense in ordinary use.' Linguistics has status in association with the human sciences, and linguists do the same sort of things that other scientists do, but the elephant in the room for this statement is the question of how linguistics relates to the natural and physical sciences. The LSA has perhaps done the best it can by not mentioning the natural and physical sciences and instead emphasizing its distinction from the humanities and preferring to talk about a scientific perspective and methods.

Linguistics has indeed emerged as a science fairly recently, among the Neogrammarians in the nineteenth century. Chapter 2 of Kretzschmar (2009) explains with some care how Saussure thought about the foundation of a science of linguistics, a discipline it its own right to be distinguished from the fields with which it overlaps. That chapter also describes how Saussure left open the possibility for a linguistics of speech, an alternative to the study of linguistic structure that he believed to be the only reasonable scientific approach for his own time. Now, a century later, we have the technology to support the study of the linguistics of speech, which Kretzschmar (2009) has identified with the study of speech as a complex system. Complexity science, however, competes today with the basic assumption of much of modern science, reductionism as the basic mode for scientific study; the basic assumptions of the two approaches are quite different. Therefore, we need to consider further than the LSA statement what it means for linguistics to be a science, including the larger picture of the epistemology and social history of modern science. A further step will be to consider science in relation to the humanities and social sciences, the first issue raised in the LSA statement, and the key to that discussion will also be the contrast between cause and effect vs. complex systems.

The epistemology of modern science: Wilson and Gould

The general problem of how we might think about cause and effect vs. complex systems, those two great ways in which the world works, came notably to the attention of the scientific world in the long-running debate between Edward O. Wilson (Figure 2.1) and Stephen Jay Gould (Figure 2.2). In his last book *The Hedgehog, the Fox, and the Magister's Pox: Mending the Gap between Science and the Humanities* (2003), before his premature death in 2002, Gould wrote about this problem in terms of the relationship between science and the humanities, and thus brought it home to observers of human behavior and culture. Gould was prompted to do so by Wilson's book *Consilience* (1998), which had argued that the humanities were just at the far end of a scale of convolution in problems that could all be solved with the scientific method. Wilson suggested, as have some linguists, that all forms of human behavior have a biological basis. Wilson's (1975) theory of "sociobiology," originally proposed in his own field of entomology, claimed epigenetic causes for what we humans often refer to as our "culture." Wilson's "consilience" located this "unity of knowledge" in the modern scientific reliance on cause-and-effect reductionism.

As Gould reports, Wilson was attracted by "the Ionian Enchantment" (Gould 2003: 191; Wilson 1998: 4–5):[1]

[1] Gould offers a long footnote about his "collegial history" with Wilson, colleagues at Harvard but not personal friends (2003: 203–205). Gould reports an incident in which protestors poured water on a defenseless Wilson (who was immobile from having his ankle in a cast), and Gould's continuing regret at not having moved fast enough himself to defend Wilson from the attack.

Figure 2.1 Edward O. Wilson (photo by Beth Maynor Young, from National Park Service www.nps.gov/aboutus/nrawards-2010.htm)

Figure 2.2 Stephen Jay Gould (photo by Kathy Chapman Online, from Wikipedia. Creative Commons 3.0 license)

a belief in the unity of the sciences – a conviction, far deeper than a mere working proposition, that the world is orderly and can be explained by a small number of natural laws. Its roots go back to Thales of Miletus, in Ionia, in the sixth century B.C. . . . The Enchantment, growing steadily more sophisticated, has dominated scientific thought.

A good example is the movement in contemporary physics to find a unified theory for all of the natural forces. Wilson (1998: 58–60), as Gould further reports (2003: 191–192), locates the basis for the Enchantment in the modern scientific mode of "reductionism"

The cutting edge of science is reductionism, the breaking apart of nature into its natural constituents . . . Behind the mere smashing of aggregates into smaller pieces lies a deeper agenda that also takes the name of reductionism: to fold the laws and principles of each level of organization into those at more general, hence more fundamental levels. Its strong form is total consilience, which holds that nature is organized by simple universal laws of physics to which all other laws and principles can eventually be reduced.

Wilson identifies "science" itself here with reductionism, the idea that we can find cause and effect, action and reaction, in smaller and smaller aspects of any problem we observe, and thus ultimately trace all the workings of the world back to physics at the subatomic level. His point of view takes the scientific method of systematic observation and makes it serve a belief in the Ionian Enchantment.

This, then, is an issue that stands behind the LSA's statement about the science of linguistics. If linguists do what other scientists do, are they also committed to reductionism? If so, Wilson has provided the connection between linguistics and the natural and physical sciences. The scientific perspective and methods that the LSA raised as a defining feature of the field tacitly join linguistics, and by implication the other human sciences, to the dominant scientific fields.

Systematic observation and the deliberate procedures of experimental science do not, however, require reductionism. We owe much of our modern scientific method to another ancient Greek, Aristotle, from the fourth century B.C., significantly later than Thales of Miletus. Aristotle began as a student at Plato's Academy in Athens, where Plato taught that the world was composed of imperfect copies of Capital I Ideas. On the other hand, Aristotle thought that, rather than begin with the Big Ideas and work down to the world, one could make systematic observations in order to build up from the *particulars* of the world to *species* (that is, specific statements that characterized the particulars), and from there up to *genus*, more general statements that characterized the species. In his *Nicomachean Ethics* Aristotle distinguished three different ways of getting at the truth: *technē*, *epistēmē*, and *phronēsis*. *Phronēsis*, or *sophia*, is most often translated as 'wisdom,' an abstract quality of mind which we all seek (in *philosophia*, as 'lover(s) of wisdom'). *Technē*, related to the word "technical," is often translated by the word *Art*, which for Aristotle is something

which enables a person to produce a certain product. The term is used not only to describe, for example, the kind of knowledge which a shoemaker needs to produce shoes, but also to describe the art of a physician which produces health, or the skill of a harpist which produces music. (Aristotle 1962: 315)

The Art of Rhetoric, to take an example from another of Aristotle's treatises, is finding the available means of persuasion in the making of an argument (Aristotle 1954).[2] *Epistēmē*, on the other hand, often translated by the word *Science*, is the result of induction from facts, such as the movement from particulars to species and then to genus, or alternatively deduction from universals:

[Scientific Knowledge] proceeds either by way of induction, or else by way of deduction. Now induction supplies a first principle or universal, deduction works from universals; therefore there are first principles from which deduction starts, which cannot be proved by deduction; therefore they are reached by induction . . . a man knows a thing scientifically when he possesses a conviction arrived at in a certain way, and when the first principles on which that conviction rests are known to him with certainty – for unless he is more certain of his first principles than of the conclusion drawn from them he will only possess the knowledge in question accidentally. (*Nic. Ethics*, 1139b; Aristotle 1934: 333)

The Art is the doing and the Science is how we can put observed facts together in order to obtain knowledge, either by induction (Aristotle's contribution) or by deduction (Plato's mode). Aristotle argues that induction is logically prior to deduction, because the first principles required to deduce something are established by induction. As Martin Ostwald suggests, "thus *technē* as an APPLIED SCIENCE concerned with production is often contrasted with *epistēmē*, which is pure scientific knowledge for its own sake" (Aristotle 1962: 315). Science, then, is first and foremost epistemology, to use another modern term from *epistēmē*, and Aristotelian science begins with induction from observations, only after which can one apply deduction from Plato's universals.

If we again consider the LSA statement for a moment, we can now see that it has not addressed *epistemē*. The LSA position makes a political statement about the association of linguistics with the human sciences, and then proposes

[2] Aristotle also gave us our foundational concepts for the study of language. His treatises called *Categories* and *Interpretation* set the model for much of modern grammar and sentence-level linguistics, and cast statements as logical propositions (Aristotle 1963). In the article in which I first discussed these issues (Kretzschmar 2006), I applied these notions of Art and Science to different modes for the study of language variation.

"formal studies" that sound more like *technē*. The statement does talk about how linguists "formulate hypotheses, catalog observations, and work to support explanatory theories," but the weight here appears to be more Platonic than Aristotelian, since the hypotheses and theories begin and end the process of analysis.

Reductionism, to return to Wilson and Gould, is a reversal of induction so that instead of taking particulars and making higher-order generalizations, one seeks ever smaller particulars out of which to make generalizations at a smaller scale. Wilson refers to the "breaking apart of nature into its natural constituents," under the assumption that smaller constituents are better ("more fundamental"), and also that we can recognize "natural constituents" when we see them after "smashing aggregates." Aristotle, on the other hand, would separate the technical act of what Wilson calls "breaking" or "smashing" of natural materials (*Technē*, *Art*) from the logical process of induction (from *Epistēmē*, Science), which provides the means to establish first principles and is therefore the heart of modern empirical science. The earlier Ionian Enchantment in its modern aspect, reductionism, may thus be seen to have more to do with Plato's Big Ideas, as a first principle that cannot be proved by deduction, indeed a principle so big that it may not be proved by induction either. Many today might assume that science is first and foremost a technical pursuit, based on the use of complicated apparatus and experimental procedures, but *technē* is just the means to a greater end. Aristotelian science relies on systematic observation of particulars (a "doing" or *Art*) in order to make generalizations by induction. Aristotelian science does not require or even license the additional practice of reduction to observe the smallest possible particulars.

Looking for the smallest components requires that the observer be able recognize the smaller particulars as "natural constituents" and not just smashed pieces, and that requires a concomitant belief that there are in fact "natural constituents" to be found – elements which together create the object as a material cause, or the effect as an efficient cause. This is exactly the problem with language study, that we cannot identify "natural constituents." Saussure has argued compellingly that we can only extract "concrete entities" from the stream of speech, elements that we can recognize as themselves even though arbitrarily extracted (see Kretzschmar 2009: 53–55). Even if we consider the physics of speech sounds, we are presented with multiple harmonics (formants) which do not present themselves as natural constituents. Just so, phonemes, the units of sound by which we tell words apart, vary in number and structure from language to language and dialect to dialect, and every phoneme in every language has a wide range of possible realizations. As we will see in chapters to come, linguists nonetheless commonly think in terms of units, or representations, as if these were the natural constituents of language. In so doing, they continue a tradition of analysis in linguistics, and also, perhaps without realizing it, align themselves with a theory of how to do science. Reductionism involves a commitment to cause and effect, and linguists who, as the LSA says, do what other scientists do, are committing themselves to assumptions that do not fit the evidence that we can collect about language.

It is perhaps ironic that Wilson bridges science and culture with reference to epigenetic rules, which "comprise the full range of inherited regularities of development in anatomy, physiology, cognition, and behavior" (1998: 163). His argument relies on reductionism, and cause and effect (1998: 292): "The strategy that works best [in various scientific pursuits] is the construction of coherent cause-and-effect explanations across levels of organization." And yet in his discussion of gene–culture coevolution, Wilson raises common cultural patterns such as aversion to snakes, avoidance of incest, and perception of the continuous spectrum of light into colors and suggests that these are the result of epigenetic rules without being able to specify particular connections of cultural patterns and genes. The consilient chain is thus broken: Wilson does not demonstrate "cause-and-effect explanations across levels of organization" from DNA to culture. Consilience cannot find its way without Ariadne's thread, Wilson's metaphor (1998: ch. 5) for movement between science and culture.

On the other hand, if we apply induction to what we have observed without reduction, without the assumption that everything we see must be some sort of "aggregate," we leave open the possibility that the whole of a thing or of an effect may be greater than the sum of its parts without an explicit causal connection – a complex system. We should not accept the idea that reductionism aptly characterizes *all* science even though it may be the dominant modern practice in science. Moreover, we can see that the Ionian Enchantment can only be a matter of belief since cause and effect may not operate in all cases, and therefore the Enchantment cannot be demonstrated by induction. We can be perfectly scientific, in the Aristotelian sense,[3] without either reductionism or Wilson's unity of knowledge, consilience. The LSA statement, therefore, appears to make a choice for reductionism by arguing for a scientific perspective, by aligning linguistics with the human sciences, and by preferring a more Platonic epistemology. Unfortunately, choosing sides in this way, even if the LSA's decision was motivated more by a common understanding of *science* than by any close analysis of what possibilities existed in the scientific world, is a mistake: it commits linguistics to a scientific model that does not fit the facts of language as we can observe them, as they were described in Chapter 1.

3 In the manner of Aristotle, though without slavish acceptance of everything that Aristotle said and did. While Aristotle is well worth remembering here for his model of scientific observation, it is of course quite true that his own observations were not always objective and unbiased, much less the conclusions he drew from them.

What it means to be scientific in the humanities and the social sciences depends crucially on just this contrast between reductionism and Aristotelian science. Wilson's consilience takes the further step of extending the Enchantment to domains in which the scientific method has not traditionally or has only recently been applied: "The central idea of the consilience world view is that all tangible phenomena, from the birth of stars to the workings of social institutions, are based on material processes that are ultimately reducible however long and tortuous the sequences, to the laws of physics" (1998: 291). While no rational person could doubt the great achievements of reductionism in many of the physical and biological sciences (Gould 2003: 216), it is, as Gould points out, another thing entirely to expropriate the scientific method and identify it with consilience, something in which one must believe. Even when there can be no effective argument, either for Gould or for the present author, against the appropriate application of cause-and-effect reductionism, still we should recognize that success in some domains does not necessarily prove success in all domains. We can doubt Wilson's exclusive belief system created by a double extrapolation from first principles, even while we maintain our confidence in the principle of scientific methods of observation. Reductionism is one "style of explanation" but it is not necessarily the only one just because it adopts the scientific method of systematic observation. And reductionism is a bad fit for language study which, as we shall see, will do better to remain in the humanities, *with* scientific methods of systematic observation but *without* the Ionian Enchantment.

An alternative to reductionism

Gould had long been an opponent of Wilson's sociobiology for another reason. He saw, *contra* Wilson, that there was indeed another process besides cause and effect by which the regularities we perceive around us could emerge. In the argument of his last book, he established the metaphor of the hedgehog, who knows only one trick for all problems (to roll into a ball, spines outward) as against the fox, who has many tricks and takes many circuitous pathways to resolve his difficulties.[4] Against the hedgehog of the dominant reductionist scientific mode, Gould set up the fox as the new field of complexity science, in which unpredictable results emerge in a process of self-organization from highly interactive complex systems. The humanities, Gould argued, are like that: culture always emerges from our social interactions but in different unpredictable ways, the product of many tricks and pathways. Gould wrote that the

[4] The magister's pox refers to the practice of censorship, related to the fox and hedgehog but less relevant here. The implication in Gould's book is that dominant belief systems color our observations of the world even when we do not realize it, a form of self-censorship. The opposition between the fox and the hedgehog comes originally from Archilocus.

cause-and-effect approach is not the best match for everything we want to study, such as the humanities, and a number of physicists, chemists, evolutionary biologists, economists, and other scientists agree with him about the importance of complexity science for their own fields, much less in the humanities.

Gould expressed the core of his argument for an alternative pathway in the importance of emergence and contingency (2003: 201–202):

> First, *emergence*, or the entry of novel explanatory rules in complex systems, laws arising from "nonlinear" or "nonadditive" interactions among constituent parts that therefore, in principle, cannot be discovered from the properties of parts considered separately . . . Second, *contingency*, or the growing importance of unique historical "accidents" that cannot, in principle, be predicted, but that remain fully accessible to factual explanation after their occurrence.

The combination of emergence and contingency constitutes the alternative pathway besides cause and effect for the way the world works. This second option does not entail the abandonment of scientific observation and procedure, or as Gould puts it, the subjects most appropriately treated in this way "remain fully within the domain of factual and knowable science." However, such subjects do entail letting go of cause and effect, because what emerges from the interaction of massive numbers of elements cannot be predicted by cause-and-effect relations between the elements even if we could know about the status and operation of each and every interaction: the whole emerges as something different from the sum of the parts. Interaction between elements makes this possible; while we might do well describing what any two elements do together as a pair, we cannot predict what those two will do in the presence of other elements with which they interact. So, while we can say with some certainty what will happen to a single grain of sand that falls onto another grain of sand (it will fall off), in a sand pile we cannot predict whether the falling grain will start a chain reaction of falling as a small landslide or as a large cascade. Random interaction makes what happens dependent on something more than the properties of individual elements.

Contingency, on the other hand, asserts the importance of the properties of individual elements and the particular conditions at the time of interaction. The randomness of interactions among massive numbers of elements in complex systems means that emergence can be different depending upon the accident of what particular elements happen to interact. Gould's example (2003: 224) contrasts the condition of quartz, which is the same whether created a billion years ago in Africa or much more recently in an atomic test in Nevada, with the sudden extinction of dinosaurs (perhaps owing to the accident of a large meteor strike) and the sudden, even meteoric rise of *Homo sapiens* within "a mere geological microsecond." Quartz is always quartz, silicon dioxide, whether in a sand pile or in larger rocks, while the complex systems of living creatures and their ecosystems, massive numbers of interactive cells or of interacting organisms, respectively, are always subject to accident, to the contingencies of time and place. A famous example of interaction with unexpected results comes from the loss of predator species in ecosystems, say, the removal of wolves. The change makes the prey species, such as deer, increase in number until the food supply fails; at that point the deer population collapses, and then enters cycles of boom and bust that are conditioned by interaction with yet other factors in the environment. The result that emerges is as bad for the prey as for the predator, even though it may appear to be a good thing for the deer to be free of wolves. And in Southeast Asia, at least at this writing, the domino theory continues to fail in its prediction that the change of politics in Vietnam would cause the entire region to make a similar change. Contingency and emergence continue among the human population in that region, and elsewhere, to create new local and regional configurations of power and social organization – including in some areas like Malaysia and Indonesia the rise of religion-based political movements, an outcome not anticipated by the framers of the domino theory who were focused on communism. Complexity theory as the alternative pathway besides cause and effect offers an opportunity for the application of the scientific method as an alternative "style of explanation" for questions that reductionism cannot answer.

Contingency is not unknown in historical linguistics. Andre Martinet (1955) considered that linguistic change could be stimulated by catastrophic events, and his arguments were taken up and discussed by Labov (1994: 21–24). Contingencies for language need not be global events like asteroids causing extinction events. Local demographic change that affects a population (like the growth in Roswell, GA, population from 15,000 in 1975 to nearly 100,000 now; see Kretzschmar et al. (2007)) might be considered to be a contingency, if not a catastrophe, that changes social patterns and thus language patterns. Indeed, local demographic changes might be considered to be a continuous series of contingencies, some minor and some major for the disruption they cause the social mechanism and thence language patterns. Contingencies related to demographic change have at best an uncertain relation to the linguistic systems, grammars, that linguists usually think about, but larger and smaller contingencies certainly affect the interactions between speakers out of which language patterns emerge. Local circumstances condition interactions such that we should expect new patterns to emerge, but as with the case of landslides and cascades in a sand pile, we will not be able to predict exactly which patterns will emerge.

Gould's second argument against reductionism as a generality deals with the incompatibility of ethical and aesthetic decisions with cause and effect (2003: 202):

> the magisterium of ethics asks a very different primary question, unaddressed (and unaddressable) by . . . factual data: What moral code *ought* we follow? What ethical

duties define a life well lived? How, in a purely logical sense, can the factual anthropology of morals resolve, or even usefully help to adjudicate, such a question?

This position addresses problems like the rise of religious factions in politics: "What moral code *ought* we follow?" The humanities and some parts of the social sciences have arisen over time as the institutional faculties that consider such questions. Neither Gould nor Wilson, however, as biologists, have an insider view of these faculties, and tend to make overly broad pronouncements about them. Wilson predicts that the humanities, though a separate domain, will be subsumed by science: "In the process the humanities, ranging from philosophy and history to moral reasoning, comparative religion, and interpretation of the arts, will draw closer to the sciences and partly fuse with them" (Wilson 1998: 12, cited in Gould 2003: 218). While an insider can make sense of Wilson's argument in some ways, in other ways it just seems patronizing, as if those who study the humanities, and for Wilson especially the social sciences, have not realized what they were doing and will in time recognize their error. There is some hint of the same indulgent view in the LSA statement about the science of linguistics. The idea that psychology, neurology, anthropology, and sociology are human sciences ignores the fact that some parts of these disciplines are more like the humanities. We should remember the difference between experimental psychology that may look like biology and abnormal psychology that may look more like the humanities, or the difference between cultural anthropology and paleontology. Even Gould is prepared to create a "magisterium of science" as opposed to a "magisterium of ethics" – later (2003: 236) "magisteria of arts and ethics (two basic domains within the 'humanities')" – as the foundation for his metaphor of the hedgehog (science) and the fox (humanities). The insider, while perhaps willing to accept that ethical and aesthetic "styles of explanation" are different from cause and effect, will likely consider that these biologists paint the humanities and the social sciences with such a broad brush that their treatment hides more than it reveals. The fifth definition of *science* in the *OED*, that the dominant modern sense in ordinary use associates *science* particularly with the natural and physical sciences, might excuse Wilson, Gould, and even the LSA to some extent, and we thus require a more developmental view to explain how it came to be.

The social history of modern science among the disciplines

The humanities and social sciences as institutional disciplines are not the effects of particular causes, but have emerged through definite historical processes.[5]

[5] Wilson argues that the arts arose among early humans "in an attempt to express and control through magic the abundance of the environment, the power of solidarity, and other forces in their lives that mattered most to survival and reproduction" (cited in Gould 2003: 240–241). This

The humanities is not just a part of the university, it is its origin. During the formative period of higher education in the West, beginning in Bologna and Paris and not long afterwards in Oxford, students first and foremost had to master Grammar, Logic, and Rhetoric. As the *OED* points out in its definition for *science*, the original sciences as recognized departments of learning were the seven sciences, or liberal arts, composed of the Trivium (Grammar, Logic, Rhetoric) and the Quadrivium (Arithmetic, Music, Geometry, Astronomy). Students had to master these sciences in Latin, for them a foreign language, and this meant learning to read and copy the great Roman writers: orators, polemicists, historians, dramatists, mythologists, fabulists, and others. There was as yet no category of writing that corresponded to our modern sense of literature, and at that time in the Middle Ages all of these writings came under the general heading of "ethics" (see Allen 1982). Early faculties of theology relied on such humanistic training to provide the literacy and rhetorical and logical skills needed for theological disputation. Gradually, through the Renaissance, the different genres of writing that we now recognize began to be distinguished, such as the separation of historical writing from fiction or mere polemic, and theology ceased to be the end towards which all scholars studied. Also at that time, in addition to writings in Latin and Greek, things written in the European vernacular languages became vehicles for study, first in Italian and French, a little later in English. This expansion of university curricula only began to include serious study of the natural sciences much later, at the end of the nineteenth century. Further great diversification of the university curriculum occurred during the twentieth century. In the early period of universities, however, language study, if not linguistics as we now understand it, was at the core of the sciences.

Now, in the twenty-first century, we have maintained the humanities as the core of the modern diversified university. A typical American university college of arts and sciences has a division of humanities, which includes departments like, at the author's university, classics, comparative literature, English, Germanic and Slavic languages, Romance languages, speech communication, history, philosophy, and religion (and thus theology has become subordinate in the university now to its former prerequisite). These subjects are all clearly related to the historical development sketched above. European universities are more diverse in their administrative structures, but humanities departments like the American ones have emerged within them. The only big change is that study of the content of writings – what we now call history or philosophy, or literature in several humanities departments – now generally enjoys a higher status than the study of the languages themselves. The other parts of the curriculum in a

sort of *ex post facto* functional argument recapitulates traditional ideas like euhemerism, which originated by recourse to imagination without much constraint in facts or history.

college of arts and sciences (or in Europe "faculties") – biological sciences, fine arts, social sciences, and physical sciences – are all johnny-come-latelys of the last century or so compared to the long history of what is now known as the humanities in the university curriculum, not to mention all of the subjects considered by other schools and colleges (or European faculties) in the modern university which arose even more recently than the arts and sciences. As it is proud to report on its website (www.ling.upenn.edu), "The Department of Linguistics at the University of Pennsylvania is the oldest modern linguistics department in the United States, founded by Zellig Harris in 1947." Linguistics as a university department is very recent indeed.

The biological and physical sciences had been studied before the late nineteenth century by individuals within institutional settings, such as Isaac Newton at Cambridge. Gould (2003: 205–215, and elsewhere) treats at length the career of William Whewell (1794–1866) at Cambridge, a teacher of Darwin. Whewell took a theological degree, then served as a professor of mineralogy for several years (and taught Darwin during that time), then became a professor of moral philosophy. Gould claims that Whewell invented the word *scientist* (2003: 206), but the *OED* (*s.v.*) gives a citation from the *Quarterly Review* in 1834 that predates Whewell's use of the term (also cited) in 1840. The late date of the first appearance of *scientist* confirms that, while today many in the biological and natural sciences may feel that their process of understanding the world is more fundamental than "soft" or "fuzzy" social sciences or the humanities, in fact well into the nineteenth century they did not even have a name for themselves. The creation of university departments in the physical and natural sciences began in earnest in Germany at the end of the nineteenth century, spurred by industrial developments like the production of modern colored dyes. Gould makes this connection of industrial applications with the natural sciences and scientists, now at the turn of the next century, with the Human Genome Project (2003: 228–232).[6] The Human Genome Project was a highly competitive commercial enterprise, with what Gould called "a powerful commercial interest" in reductionism.[7] The Ionian Enchantment with unity of science, therefore, was jump-started in the late-nineteenth century and confirmed with the passage of time as reductionism, at least in part because of its commercial connections. While it is certainly true that Aristotle and the Greeks launched the scientific

[6] At the end of the race, it has been found that a total of about 30,000 genes are used to produce about 150,000 proteins – more evidence for emergence and contingency, since only a process of interaction could produce five times as many proteins as there are genes.

[7] "If a disorder arises from a particular misconstruction in a specific protein, and if we could sequence the gene coding for this protein, then we might learn how to 'fix' the gene, correct the protein, and cure the disease. Thus the debate about patenting genes represented no mere academic exercise for a university's moot court, but rather reflected a driving commercial concern of the large, growing, and highly speculative industry of biotechnology" Gould (2003: 229).

method of systematic observation and the use of logic in the Classical Age, the boat that is modern reductionist science actually left the harbor only in the late-nineteenth century, having been funded by investors from commerce. The history of science has its own sociology.

The humanities cannot exist in an ivory tower (no matter the popular metaphor) any more than the physical and biological sciences, and so we must also consider the social development of the humanities. During the formative period of Western education, the humanities taught university graduates to become the literati who served an illiterate aristocracy in secular affairs and an unlettered public in religion. Gradually, over the centuries, the sons of men of means attended the university themselves, and so over time a university education became the hallmark of a gentleman, not just of the gentleman's clerk. And in addition to the technical aspects of the discipline – grammar, rhetoric, and logic – the humanities offered these young gentlemen a system of social values, nothing short of our common Classical Western inheritance, including the ideals of democracy, of justice, of tolerance, and of the necessity in society for citizens with greater means to show liberality to their fellow citizens. The writings that the Middle Ages had all classified as ethics still taught ethics to these young gentlemen, now from the point of view of history, of literature, and of philosophy. Indeed, as John Henry Cardinal Newman argued in 1852, the ideal of a liberal education in the humanities had become inseparable from the ideal of a gentleman (1990).

Furthermore, education in the humanities became the sine qua non of high social status. Knowledge of the humanities was worn as a badge of honor by gentlemen, so much so that their journals and letters were peppered with classical allusions. The failure of the nouveau riche to cite the Greats (as the Classics were called at that greatest bastion of gentility, the University of Oxford) was held against them by real gentlemen as a lack of class, a social shibboleth. Conversely, some ability to cite the Greats could confer respectability upon the user: even the worst scandal sheets among nineteenth-century American newspapers show a high frequency of classical allusions (see, for instance, sensational accounts of a period murder presented in Bailey 2003). The humanities, therefore, was no longer just an academic discipline. At the same time that the humanities embodied the central social values of Western culture, knowledge of the humanities came to be employed by gentlemen as a social discriminator. The humanities had thus developed over time within the university from language study of writings that expressed man's best qualities and better nature, to become an important measure of men in society. I have been saying men here, not men and women, because women were routinely excluded from the university at that time, something we have of course changed for the better, increasingly in the physical and biological sciences as well as in the humanities.

The status of the humanities thus differs from the institutional status of the fine arts and the social sciences. The creative arts have of course been practiced ever since there have been human settlements (in the remains of which we find evidence of early human artistic production, such as cave painting). An ability to create graphic or plastic images, or to play music, or to enact dramatic presentations, has not enjoyed high social status for very long in comparison with the trappings of the humanities for gentlemen, indeed quite the opposite. As the nineteenth century progressed, women, still largely excluded from universities, were expected in polite social circles to be able to play an instrument or to take an interest in literature or painting, skills that men need not cultivate because of their involvement in more serious aspects of the humanities and in the commercial world. The inclusion of the fine arts in university institutional settings is still complicated by the existence of separate conservatories for music or schools of art and design, not to mention the various means to teach and learn acting.

The social sciences emerged within universities at roughly the same time as departments in the physical and biological sciences. They, too, were quite separate from the long social development of the humanities, and their claim for new status as reputable departments in the university was parallel to the claim of the physical and natural sciences. Anthropology, psychology, and sociology have indeed had their successes as they have tried to apply scientific methods to the study of human culture – just not reductionism, unless in the more biological approaches to these disciplines. Indeed, one of the most effective movements in sociology (Durkheim), anthropology (Lévi Strauss), and linguistics (Saussure) was in the other direction, towards structuralism that emphasized relationships within social organization understood in itself as an object (a "social fact"). Abstraction away from individual behavior and individual interaction, which until recently in these fields have not been tractable for study in themselves except as case studies, allows the researcher to make headway by resort to a higher level of analysis, not by reduction to ever smaller levels. In each of the social sciences there remains a tension between more "scientific" (usually biological) approaches and more humanistic approaches, and Wilson predicts their eventual separation. In linguistics, generativism would fall into his "scientific" camp with its biological associations, while more humanistic, cultural approaches would go in the other direction. Again, the social history of the fine arts and social sciences as institutional disciplines strongly conditions their status, both in relation to the physical and biological sciences and to the humanities.

The differences in the histories of the humanities, the fine arts, and the social and natural sciences all come together when we consider the study of language. Kretzschmar (2009) considers in detail how the institutional study of linguistics emerged from the humanities and sought its own space in contrast to notice taken of language in the various social sciences (Chapter 2). In the creation of late-nineteenth and twentieth-century linguistic theory, Saussure and the Neogrammarians chose mechanical, cause-and-effect processes within linguistic structures in order to counter earlier unsystematic opinions of language. On the other hand, we have the peculiar status of English teachers (including the present author) as representatives to the public of the humanities. Since the humanities has become the measure of social status, English teachers have become social gatekeepers. Some English teachers may even have entered the field to serve as conservators of gentlemanly society, or for language specialists, to be the guardians of linguistic propriety. Whether or not we want that gatekeeper job, we have it anyway because the public gives it to us. Generativism, as a social science, is no help here against the social power vested in teachers of English, as a bastion of the humanities. At the end of the day, the humanities has a special status both within the institution and in the popular view that has little to do with the scientific method or reductionism, and much to do with the way that we interact in our cultural complex systems. When the LSA talked in its statement about a "purely cultural academic discipline in the humanities" from which linguistics had now emerged, it distanced linguistics as a science from such social entanglements.

This social history is not what Wilson or even Gould was talking about as the humanities, which they conceived, respectively, as being dominated by the creative arts, and being its own ethical and artistic magisterium. The insider view offers a richer, more nuanced view of the humanities as it continues to emerge in social interaction. Nowhere in the social history of academic disciplines is there any reason to think that researchers in the humanities and social sciences cannot use systematic methods of observation, a scientific perspective in the Aristotelian sense. Similarly, there is good reason to think that commercial interests have had an important role in the more recent development of the natural and physical sciences. Language study has been a recognized science from the beginning of universities, and it is only recently that linguistics has, as shown in the LSA statement, distanced itself from the humanities as a "purely cultural" institution. In so doing, linguistics has thrown the baby out with the bathwater: the humanities can be scientific just as well as the natural and physical sciences can, and research in the humanities has the advantage of working with human populations both historically and in the present day. Thus the movement in the LSA statement from the former position of language study in the humanities to a modern linguistics that overlaps the human sciences is really a false distinction from a scientific perspective. The change is mainly political, based on social factors, and in its association with the Ionian Enchantment it leads linguistics away from systematic observation of the evidence that we can collect about language.

The humanities as modern science?

Some humanists, too, have sought to connect the humanities with modern science. Gottschall's *Literature, Science, and a New Humanities* (2008) argues that "the field of literary study is in bad shape," and that "there is no real epistemological divide between the sciences and the humanities" so that "there is nothing but inertia stopping literary scholars from adding powerful scientific tools to their existing analytical toolkits" (Gottschall 2010: 301). His book offers quantitative analyses of folktales to validate "biosocial versus strictly social-constructivist theories about gender, literary universals, and romantic love" (2010: 302). The argument of this chapter does not support Gottschall's assertion of consilience between science and the humanities, and suggests that the addition of scientific tools like statistics to humanistic toolkits would not in itself constitute science.[8] Application of modern reductionist science may be possible to some degree but, if we accept Gould's arguments, the humanities is better characterized scientifically by complex systems. The central problem for Gottschall is the social history of and competition between the disciplines, in his words, that "the sciences are thriving in all of the dimensions – funding, morale, prestige, intellectual momentum – in which the humanities are sputtering. If we in the humanities hope to halt and reverse this process of decline perhaps we should begin by studying the secrets to the sciences' success" (2010: 302). Again, the argument of this chapter does not support the idea that putting the sheep in wolf's clothing will address the real social problem of the humanities, that it has not been connected with commercial interests to the same degree as medicine and the natural and physical sciences. Deborah Cameron, in a response to Gottschall, makes just this point (2011: 70):

> I do not see much evidence that literary studies lacks purpose or confidence. But if it did, I doubt whether the solution would be for it to attempt to rebrand itself as a branch of the human or cognitive sciences. For one thing, it seems highly unlikely that this strategy would reap the rewards Gottschall hopes for, like more generous funding or higher esteem . . . For another thing, I think the risk is considerable that in trying to become more 'scientific', literature scholars will produce work that is second-rate by the standards of both literature and science.

Cameron's risk here surely stems from the ethical and aesthetic concerns of the humanities that modern science is not well equipped to address. Cameron also warns that abandonment of the historical pluralism of study in language and linguistics is dangerous, i.e. giving up the application of methods from both the humanities and the social sciences in favor of any single approach. As she says, "which approaches or methods are most fruitful for linguists' purposes depends on what aspect of language they are trying to understand" (2011: 70). Cameron, then, might stand with Gould, and Gottschall with Wilson, as these

[8] So, for example, application of formulas from nonlinear mathematics, fractals, have been used to create graphic designs used in fabrics for clothing; repeating patterns from cellular automata have likewise been used for purely aesthetic purposes, and programs are available on the Internet to create designs with "Rule 30" and other algorithms derived from Wolfram's (2002) comprehensive study of one-dimensional automata.

humanist insiders evaluate their relationship with the sciences, in view of both the epistemology and the social history of the disciplines.

The humanities does not have to give up the nature of the discipline as it applies new methods including new technology to the study of language and literature. Digital humanities has become a strong subfield, with international organizations (Association of Digital Humanities Organizations, Association for Computers and the Humanities, European Association for Digital Humanities, among others) to advance the work and promote it. Indeed, the central journal in the field, *Literary and Linguistic Computing* (recently renamed *Digital Scholarship in the Humanities*), explicitly connected linguistics with the humanities even in its title, and the journal continues a strong tradition of language study in areas not well covered in American linguistics departments, such as corpus linguistics and authorship attribution. Gould suggests that, since "factual information in scientific form will be extremely helpful and relevant to the discussion of almost any important question" in the humanities, "any humanist who would reject this aid, this helpful hand of friendship, is either a narrow pedant or a fool" (2003: 235). To interpret Gould's statement as saying that humanists should have to struggle to understand factual information misses the point; humanists just need to become familiar with the scientific form in which factual information may be conveyed. This is one of Gould's strong points, the conveyance of information to a wider public that, in many cases, other scientists' lack of familiarity with the humanities has prevented them from sharing. The humanities and the more humanistic parts of the social sciences now particularly embrace the role of looking for facts and new styles of explanation to apply to the traditional questions of the humanities and human cultural interaction. Of course this does not mean that humanists and humanistic social scientists want to abandon the humanities in favor of modern science, much less to merge the humanities with reductionism under the Ionian Enchantment. Rather, as Gould also suggests, "useful collaboration between basically different entities with strong common interests does not imply fusion within a basically linear hierarchy of common structure" (2003: 235). Gould's great contribution to "mending the gap between science and the humanities" has been to show that science is no longer as unified under reductionism as Wilson's *Consilience* had argued, and to suggest that the new scientific form that parallels cause-and-effect reductionism, complex systems, requires new evaluation as a bridge between science and the humanities. The new humanities must still work with ethical and aesthetic questions that cannot be addressed with reductionist methods. Ethical and aesthetic questions have no cause-and-effect answers; at any given time the ethical and aesthetic views of culture are the product of interaction between the members of the culture, always allowing for contingencies. Of course, different cultures can and do have different ethical and aesthetic values, and always have had; moreover, the ethical and aesthetic views in a culture are subject to change, perhaps given new contingencies or maybe just as the result of continuing interaction. The real division in modern science arises from the essential difference between reductionism and complexity science, and this difference cuts across traditional institutional divisions between the natural and physical sciences, the social sciences, the arts, and the humanities. For the LSA to remove linguistics from the humanities and to choose an association with just one approach to science thus marginalizes the discipline by making it turn away from a genuine scientific perspective that fits our evidence about language and culture in order to become a kind of camp follower for the physical and natural sciences.

3 Usage-based linguistics and complex systems

The American novelist Walker Percy was not a trained linguist, but he had a passion for language and a strong interest in linguistics. In 1977 he commented on what he saw as the state of linguistics, framed here in the traditional terms of rational vs. empirical approaches:[1]

> Scientists [of language] tend to be dogmatic about the nature of man. Again they remind me of the Scholastics battling with Galileo. Scholastics spent thousands of man-hours inside their heads trying to prove that Jupiter couldn't have moons and that the earth was at the center of the universe. To suggest otherwise offended their sense of the order of things. Galileo pointed to his telescope: Why don't you take a look? Today we have plenty Scholastics of language. What we need is a Galileo who is willing to take a look at it.

When Percy died in 1990 the movement in linguistics that followed his advice, usage-based linguistics, was just getting started. Paul Hopper, for example, presented his seminal paper called "Emergent Grammar" in Berkeley in 1987. I would like to trace the development of usage-based linguistics and its association with complex systems by touring some landmark North American works in the field. Usage-based linguistics should represent the closest, most empirical approach to speech now obtaining in the field of linguistics.

The point of this exercise is to show that work by usage-based linguists and study of complex systems in speech, while making great progress, continue to be handicapped by notions carried over from Percy's "Scholastics of language." There is some irony in the fact that many of the authors discussed here argue quite sharply against the generative model as they attempt to defend their usage-based model, and yet they still strive to align what they do with ideas derived

[1] The reference to Galileo is not uncommon, and is in fact used by linguists on both sides of usage-based study (see Kretzschmar 2006). I am grateful to my student Calen Verbist for making me aware of Walker Percy's ideas about linguistics. I am also grateful to members of the Workshop on Cross-Linguistic and Language-Internal Variation in Text and Speech, Freiburg, 2011, for many useful comments. The works on usage-based linguistics treated in this chapter come from the American tradition in this area, and in Europe usage-based studies have a different trajectory. For the development of the "language in use" model in Britain, see Kretzschmar (2009: chs. 1 and 5). This chapter is based on Kretzschmar (2014).

from formal linguistics. Usage-based linguistics simply cannot use the same model as generative linguistics, especially when we apply complexity theory in our work. As will become clear, the formal notion that a language "has" some particular feature in it, and therefore that it "has" a particular aggregation of features overall, is greatly complicated – indeed becomes untenable – once we pay attention to the effects of complex systems on language in use. We will see that a complete single grammar for a language could never be motivated by detailed observation of speech production, because the patterned distribution of features and their variants that always emerges from the complex system of speech cannot be captured by the binary logic of formal linguistic systems. To be true to itself, usage-based linguistics should discuss the usage-based patterns in language that actually emerge from its complexity, whether for individual features or in the aggregate, and allow others to derive the benefits that arise from formal analyses.

Paul Hopper (1987)

Let us begin the tour in the early days for usage-based linguistics and complex systems, with Paul Hopper and his influential 1987 paper, "Emergent Grammar." His position accords exactly with complex systems:

> This is, then, roughly the context in which the term Emergent Grammar is being proposed. The term 'emergent' itself I take from an essay by the cultural anthropologist James Clifford, but I have transferred it from its original context of 'culture' to that of 'grammar'. Clifford remarks that 'Culture is temporal, emergent, and disputed' . . . I believe the same is true of grammar, which like speech itself must be viewed as a real-time, social phenomenon, and therefore is temporal; its structure is always deferred, always in a process but never arriving, and therefore emergent. (Hopper 1987)

Hopper borrowed the term "emergence" from cultural anthropology, not complex systems, an example of how similar ideas can be "in the air" at a moment in time. If the structure of grammar is "always in a process but never arriving," a wonderful phrase, this means that we can conceive of grammar as a rational object, but one that is never instantiated, always only indirectly related to speech. Hopper explains further:

> The notion of emergence is a pregnant one. It is not intended to be a standard sense of origins or genealogy, not a historical question of 'how' the grammar came to be the way it 'is,' but instead it takes the adjective emergent seriously as a continual movement towards structure, a postponement or 'deferral' of structure, a view of structure as always provisional, always negotiable, and in fact as epiphenomenal, that is at least as much an effect as a cause.

Grammaticalization, as we now call this process, is not an explanation for current structure in language, but instead should be understood as "continual movement." Structure is "epiphenomenal" not because it is unimportant, but because it is always contingent and never directly observable. Finally, Hopper affirms that "because grammar is always emergent but never present, it could be said that it never exists as such, but is always coming into being. There is, in other words, no 'grammar' but only 'grammaticization' – movements towards structure which are often characterizable in typical ways." I would modify this striking statement only to clarify that, while grammar never exists as such in language in use, it can well exist as a description of regularities indirectly derived from speech performance by perceptual means. This is, in fact, just what all linguists do. Grammar is not generative in the sense that its basic patterns control or motivate what people say or write; a grammar as linguists create it is instead an *ex post facto* generalization that more or less describes what some person or group of people have said or written. The next chapter addresses this point in depth.

If we observe the distributional pattern of linguistic features at any moment in time, we can get more historical information than we might have thought. Hopper (1987) used the example of the development of the English indefinite article *a/an:*

> To take just these three functions of the predecessor of *a/an* in Old English, we find in modern English not a uniform, over-all weakening of the meaning, but rather a situation in which the weakened meanings and the older stronger meanings exist side-by-side
>
> They introduced me to a young woman [whose name was Ethel]; My husband and I went to a showroom to pick up a new car we had ordered. After we had test-driven it, the salesman asked us . . .
>
> VS.
>
> Birds of a feather flock together; They are all of a kind; A stitch in time saves nine; A penny saved is a penny earned; How much is that picture-frame? – A dollar; Linguists of a theory attend the same conferences; What was left of the woods after they built the parking-lot? – A tree.

In other words, English has preserved the historical functions of *a/an*, as well as developing new ones. The *OED* entry for *a/an* as updated for the third edition describes what Hopper illustrates in three different levels of definiteness with a total of twelve subheadings, plus a sense not shown by Hopper indicating proportion (*miles an hour*, *dollars a week*). Whether in proverbial language or in special domains like law or religion, we do still employ many old constructions of the kind that Hopper offers, which gives them a low frequency on the modern A-curve. The indefinite article is not an exception but a good example of a general capacity of speech to retain old feature variants at low frequencies. Stockwell and Minkova (1988), for instance, reported that it was possible to find vowels at every stage of the Great Vowel Shift in modern use somewhere in the

Survey of English Dialects. Similarly, Allison Burkette (2001) has shown that in the modern A-curve of different words used for *bureau*, as constructed from the LAMSAS survey of American English, low frequency variants in the long tail are often names for that piece of furniture used in Colonial pattern books for furniture makers. In historical semantics, one of the points made by C. S. Lewis in his *Studies in Words* (1960) is "ramification," the fact that words do retain their older meanings even while they gather new ones. Thus, for articles, vowels, the lexicon, and semantics, all of the evidence suggests that historical forms tend to be retained as low-frequency variants in the tail of the nonlinear distribution of contemporary usage. By the same token, the long tail is where we will find novel features as they enter the language. Grammaticalization, then, is not a process in which particular variants are selected as being "grammatical," but instead it is the process by which variants just become more frequent at the expense of older variants, and then themselves may become less frequent as newer variants arise and expand. What we call grammaticalization thus corresponds to change in frequency on the A-curve, and in particular for some usage its increase in frequency on the curve as it becomes one of the few top-ranked features.

Hopper's emergence thus corresponds exactly to what happens in complex systems. Hopper did not describe how frequency profiles emerge, and he did not describe the scale-free nature of complex systems. However, he did grasp and present the essential dynamic movement of variation in speech, always in a process but never arriving, and he presented evidence that supports the idea of the A-curve.

Paul Hopper and Elizabeth Traugott (1993)

The next stop on the tour again features Paul Hopper, this time with Elizabeth Traugott in their 1993 book *Grammaticalization*. They define their key term as follows: "We define grammaticalization as the process whereby lexical items and constructions come in certain linguistic contexts to serve grammatical functions, and, once grammaticalized, continue to develop new grammatical functions. It is the process whereby the properties that distinguish sentences from vocabulary come into being diachronically or are organized synchronically" (Hopper and Traugott 1993: xv).

This statement cannot be more different from the 1987 article. Grammaticalization is still a process, but it is defined by how the properties of sentences "come into being" or have synchronic organization. Grammar has become an object, no longer a process. Further to this point they say:

Grammaticalization . . . refers to that part of the study of language that focuses on how grammatical forms and constructions arise, how they are used, and how they shape

> the language. The framework of grammaticalization is concerned with the question of whether boundaries between categories are discrete, and with the interdependence of structure and use, of the fixed and the less fixed in language. It therefore highlights the tension between relatively unconstrained lexical structure and more constrained syntactic, morphosyntactic, and morphological structure. It provides the conceptual context for a principled account of the relative indeterminacy in language and of the basic non-discreteness of categories. (Hopper and Traugott 1993: 1)

So, now grammaticalization exists with reference to a framework, a set of categories which are non-discrete. Items may be "more" or "less" grammatical, not just more or less frequent. Grammar is contingent (because the categories are non-discrete, not naturally given as discrete), but nonetheless objectified, reified, and above all characterized by constraints of "syntactic, morphosyntactic, and morphological structure." Hopper and Traugott elsewhere use nominalizations like "reanalysis" (e.g., in chapter 3, on "reanalysis and analogy") and passive constructions like "have been generalized, or analogized," which leave out the agent for the action. The implication of such phrasing is that users of language are the agents, but really the agents doing the reanalyzing, generalizing, and analogizing are the linguists themselves as they create a grammar. What may have begun as a usage-based argument has now become highly formal, so that Hopper and Traugott treat the hypothesis of unidirectionality cross-linguistically and make it a "metacondition on how particular grammatical constructions will change" (1993: 95). So, in *Grammaticalization* the book, what had started in "Emergent Grammar" the article has become radically more aligned with formal linguistics. The second edition of the book does mention emergence, in one line of a new paragraph on usage-based structure (2003: 35), but does not alter the thrust of the book. Hopper's brilliant insight in the article, the idea of grammar as continual movement, has been returned to a much more mainstream formal discussion in both editions of the book. Hopper has explained to me (p.c.) that he had not actually changed his mind about emergence, but that he needed to accommodate his coauthor in order to pursue the joint project, and that meant moving the discussion back towards the mainstream. Owing to the great success of the book, Hopper's original ideas about emergence then receded from view and linguistics backed away from complex systems.

For usage-based linguistics, this represents a major retreat from the central idea of emergence as a result of a decision motivated by academic polity. Hopper's decision to collaborate with Traugott impeded progress in linguistics towards complex systems that could have begun in the 1990s, as the Sante Fe Institute promoted complexity science in many other fields. We should get along with our colleagues, certainly. But Hopper made a mistake and, ironically, did harm to our field when he backed away from the implications of his striking new insight about emergence in order to fit in better with other linguists.

Joan Bybee (2001)

Another stop on the tour, Joan Bybee's 2001 *Phonology and Language Use*, cites Hopper's article but does not cite Hopper and Traugott. Nonetheless, she describes grammaticalization much as it is presented in Hopper and Traugott's book. The first and last lines tell the story:

> The present work proposes to demonstrate that the focus on structure needs to be supplemented with a perspective that includes more than just structure, a view that includes two other important aspects of the language phenomenon – the material content or substance of language, and language use. The substance of language refers to the two polar ends – phonetics and semantics – that language molds and structures, the two ends between which language forms the bridge. Language USE includes not just the processing of language, but all the social and interactional uses to which language is put. For present purposes, in the context of phonology, the frequency with which certain words, phrases, or patterns, are used will be shown to have an impact on phonological structure. (Bybee 2001: 2)

Bybee's focus remains on abstract structure, which she will "supplement" with information from phonetics, semantics, and language use. The latter, in the form of the relative frequency of particular words, phrases, or patterns, will merely have an "impact" on phonological structure. Thus phonological structure is a given, an assumption, that usage can only enhance or affect. Bybee (2001: 19) later specifically aligns her discussion with the goals of generative phonology. From such a beginning, it is difficult to imagine how much enhancement, or what sort of impact, language in use can actually have in a usage-based linguistics.

Bybee is the first on the tour to address complex systems, and wants to see "segments as emergent units" (2001: 85):

> In network models, internal structure is emergent – it is based on the network of connections built up among stored units. The stored units are pronounceable linguistic forms – words or phrases stored as clusters of surface variants organized into clusters of related words... Units such as syllables and segments emerge from the inherent nature of the organization of gestures for articulation.

Bybee suggests that items from the stream of speech, pronounceable linguistic forms, are retained in memory, and that the storage of many items leads to "clusters" at a higher level of organization. Then further, Bybee claims that (2001: 86)

> The view that phonological representations are self-organizing means that units of analysis, such as segments and syllables, are emergent units and are permitted to have gradient properties. This view does not insist upon one unit of uniform size for describing all speech, but rather proposes that the organization of linguistic material into units depends entirely upon the substantive properties of that material.

Self-organization, another key term from complex systems, thus leads in Bybee's view to the instantiation in the brain of linguistic structure, units like segments and syllables that linguists manipulate. The problem is that her a priori assumption of linguistic structure means that Bybee has prejudged the outcome of emergence. As we have seen in Chapter 1, every individual is at the nexus of many groups of people, regional and social, and each one of those groups will have its own distributional pattern of many variants for any feature we name.

Figure 3.1 provides a good example of this fact in an acoustic phonetic plot showing variant realizations of the /u/ vowel in a group of young Georgia speakers. Each student carried out a fixed format elicitation task, in which each of seventy-eight cues (single words) was recorded twice. These cues cover the vowels usually described for American English, each one in several environments. For the vowel characteristics, fundamental frequency (F0), F1, F2, and F3 of each token were measured from the display of the TF32 program implemented in CSpeech SP (Milenkovic and Read, 2000). For monophthongs, the frequencies are measured in the center of the vowel segment. In Figure 3.1, among sixteen students at the University of Georgia, we can count the number of tokens in each cell of a grid overlaid on acoustic F1/F2 space (adapted from Kretzschmar, Dunn and Kim 2011, shown in grayscale). The frequency ranking overall for the count of realizations per cell of the grid is shown in the A-curve at bottom right. We can see not only that the men (circles) are different from the women (triangles), but that the different environments for the sound (e.g. the distinct cluster of *pool* variants at the right of the chart, cue 3) each have their own distributions as well. We can create an overall generalization for the phoneme /u/ in this group of speakers, or separate generalizations for men and women – the M and F labels on Figure 3.1 indicate the average F1/F2 coordinates for the two groups, and the overall average would be between those two points – but we could choose to emphasize the different environments instead. There is no necessity to privilege the units that linguists normally talk about, like phonemes, which derive essentially from the simplifying generalization of a formal model. Similarly, the point of Kretzschmar (2012: 389) moves in the same direction:

> we can see why we might be led [by the phonetic evidence from LAMSAS at Kurath's proposed North Midland/South Midland dialect boundary] to make qualitative statements – the top-ranked variants are much more common than most others – but we need to understand that we can only use the notion of "contrast" to make the grossest of generalizations about essentially arbitrary divisions between arbitrary categories for different realizations. We see clearly from our quantitative analysis the important fact that features of speech are actually continuously variable both in word classes and in regional and social groups.

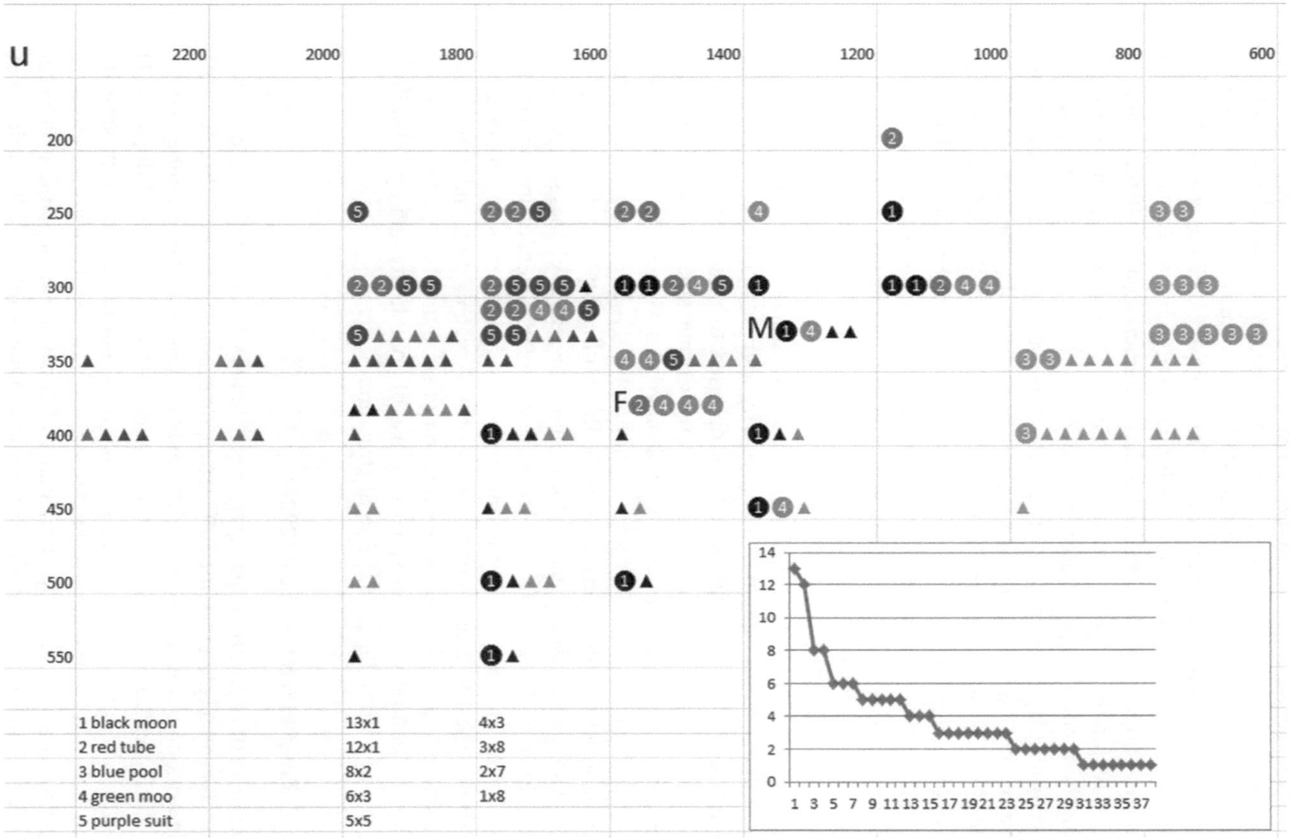

Figure 3.1 Distribution of UGA /u/ realizations (148 tokens, 5 environments). Circles represent male speakers, triangles female speakers; numbers in the circles indicate which word yields the token for the men, shown only by shades of gray for the women.

Inspection of the phonetic record for five different vowels among the speakers surveyed near the boundary shows substantial overlap between the realizations for the proposed North Midland and South Midland areas, among a much wider range of vowel realizations than anyone might have expected, and so the idea that the two regions contrast with each other on the basis of that evidence is difficult to support. How to make the best characterization of the range of variation within word classes and different groups of speakers is thus an excellent question, not a simple matter of applying a set of phonemic categories.

To return to Bybee, her main contribution to usage-based linguistics is her suggestion that the frequency of words contributes to the construction of the phonology of a language (first introduced in Hooper 1976). When she talks about frequency in Bybee (2001), she most often refers to Francis and Kučera's *Frequency Analysis of English Usage* (1982), derived from the Brown Corpus of written material from the 1960s. While the use of Francis and Kučera does offer her support for her thesis about the importance of frequency, her leap to top-level analysis of English conceals the kind of variation on the ground that we have seen for the Georgia /u/ vowel and for realizations of vowels at the North Midland–South Midland boundary. Her argument about the emergence of cognitive structure in individuals, as opposed to effects in the language overall, cannot fairly be related to the Brown Corpus since no individual, not even Francis or Kučera, has ever all and only been exposed to speech at the same frequencies of variants as occur in the corpus. Bybee's a priori assumption of formal structure thus leads her to take the complexity out of the complex system. Application of frequency information at the top level does succeed, as it ought to do given the scale-free nature of the complex system of speech, but the top level generalization is not the only one we should recognize. In her 2001 book Bybee tried to marry simple ideas of self-organization and emergence from complex systems with the assumption of structure, but she was doomed to fail, if influentially, by her superficial use of complexity and by her failure to break away from the assumption of structure.

Janet Pierrehumbert (2001)

The problem of appropriate management of frequency information is addressed by Janet Pierrehumbert in a fine 2001 article on exemplar theory. While Bybee's earlier work on frequency noticed differences in corpora, Pierrehumbert (starting from Bybee's insight as shown in Hooper 1976, and from Phillips 1984) transferred the focus to individuals, to memory and the cognitive processing of speech forms. Exemplars as a model for perception and categorization were introduced to psychology by Hintzman and Nosofsky in separate articles in 1986. Keith Johnson had introduced the idea to linguistics in a 1997 article, but Pierrehumbert's exposition is the one commonly cited and discussed.

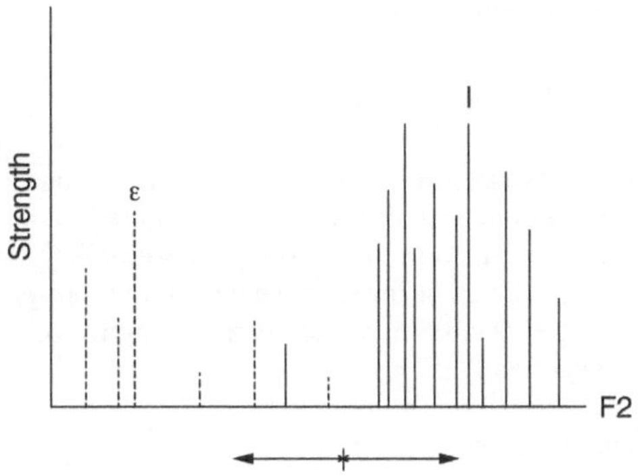

Figure 3.2 Pierrehumbert decision model (from "Exemplar Dynamics: Word Frequency, Lenition and Contrast." In J. Bybee and P. Hopper, eds., *Frequency and the Emergence of Linguistic Structure* (Amsterdam: John Benjamins, 2001), 137–157. Reprinted with kind permission from John Benjamins Publishing Company, Amsterdam/Philadelphia, PA, www.benjamins.com)

Pierrehumbert clearly distinguishes exemplar theory from "standard" models of phonetics and phonology (2001: 138):

> First, in all standard models, the lexicon is distinguished from the phonological grammar. The exact phonetic details of a word's pronunciation arise because the word is retrieved from the lexicon, and processed by the rules of constraints of the grammar whose result (the surface phonological form of the word) is fed to a phonetic implementation component [which] computes the articulatory and/or acoustic goals which actualize the word as speech . . . A second challenge arises from the fact that differential phonetic outcomes relate specifically to word frequency. Standard generative models do not encode word frequency.

On the other hand, in the exemplar model, "each category is represented in memory by a large cloud of remembered tokens of that category" (2001: 139). The task for anyone listening to a speaker, then, is to hear a new token and compare it to the distribution of remembered tokens in order to make the best decision on what category the new token belongs to. The process by which the decision might be made is illustrated in Figure 3.2. The new token is symbolized by the asterisk at the bottom of the figure, and the task is to decide whether the token belongs to the category for the /ɛ/ phoneme or for the /ɪ/ phoneme along the one dimension presented here, F2. In this ambiguous case, the new token is closer to the main distribution of /ɪ/ remembered tokens, even though the actual distribution overlaps, and so it would be assigned to the /ɪ/ category. Pierrehumbert notes that the decision must necessarily be affected by what the ear can distinguish, and by how recently each exemplar was encountered. She also explicitly allows that any exemplar can be a member of more than one classification scheme at the same time, so that one exemplar might be included in classifications for a phoneme, a word, female speech, and "Mom" all at the same time (140). As Pierrehumbert neatly summarizes (142):

> The exemplar approach associates with each category of a system a cloud of detailed perceptual memories. The memories are granularized as a function of the acuity of the perceptual system (and possibly as a function of additional factors). Frequency is not overtly encoded in the model. Instead, it is intrinsic to the cognitive representations for the categories. More frequent categories have more exemplars and more highly activated exemplars than less frequent categories.

The particular benefits of the exemplar model, then, are the improved description of detailed phonetic knowledge that speakers have about their language, the mechanism for decisions about categories, and the accommodation of frequency to some degree.

These are significant benefits, and yet the model does not move as far as it might have away from "standard" models and towards complex systems. First, the notion of category, also called "label" by Pierrehumbert, is much simplified in Figure 3.2. The data for *six* from the Linguistic Atlas in Chapter 1 has no examples of transcriptions with /ɛ/. However, in the Atlas question that elicited the phrase *sit down*, 592 people answered *sit down* and 497 people answered *set down*. In such a case any decision about phoneme categories may be confounded by the alternation in vowels. Similarly, the Atlas data for *ten* as shown in Table 3.1 has 29 transcription types with /ɪ/ and 30 transcription types with /ɛ/. Perhaps the total of 59 types could be reduced, with many exemplars merged if these fine transcriptions were judged to be below the acuity of many hearers (even if they could be and were heard by the LAMSAS professional transcribers). Even so, the comparison of any new token to remembered exemplars will certainly be much more complicated than Pierrehumbert's model suggests. Table 3.1 also points out another flaw in the simple exemplar model: the frequencies shown in Figure 3.2 are not in the A-curve pattern that we expect of a complex system. The vowel in *ten* in the Atlas data has 29 transcription types with /ɪ/, but none of them are highly ranked. The top six types, all with /ɛ/, account for about 75 per cent of the tokens (897/1192). Surely the massive advantage of top-ranked variants that we expect to see from nonlinear distributions of tokens should have an effect on decisions about fitting exemplars into categories.

Another challenge to the exemplar model is statistical learning in language acquisition. In a landmark article, Saffran, Aslin, and Newport (1996) reported that 8-month-old infants could learn to distinguish "pseudoword" cues composed of three syllables from "nonword" cues composed of the same syllables in a different order, after just two minutes of training with the pseudoword cues. In order for this to occur, the infants must have been able to learn from their exposure to the order in which the syllables were most likely to occur. This research

Table 3.1 *Transcriptions of* ten *from LAMSAS*

tɛ{ə}n	187	tɪ∧·{ə}n	2
tɛ∧{ə}n	161	tɨn	2
tɛ∧·{ə}n	158	tɪ̃∨n	2
tɛ∧n	149	tɛ̆>n	2
tɛn	140	t∨·{ }n	2
tɛ·{ə}n	102	tɛ∨·{ə}n	2
tɪ·{ə}n	33	tɛ∧{ɨ}n	1
tɛ>n	33	tɪ{ɨ}n	1
tɪ∨{ə}n	24	tɪ∧{ɨ}n	1
tɛ∧·n	19	tɪ∨{ɪ}n	1
tɪn	19	tɪ∨·{ɨ}n	1
tɛ·n	16	tɪ∨>n	1
tɪ·{ɨ}n	15	tɛ>{ə>}n	1
tɪ{ə}n	14	tɪ̃∨{ə̃}n	1
tɪ∨n	14	tɛ̆{ə}n	1
tɪ∨·{ə}n	10	tɛ∧{ə∧}n	1
tɛ>{ə}n	9	tɛ̃∧{ə̃}n	1
tɪ∧{ə}n	6	tɛ∧<{ə}n	1
tɪ·n	6	tɛ∧>{ə}n	1
tɪ∧n	6	tɛ̃·{ə}n	1
tɛ̃∧·{ə̃}n	6	tɛ̃·{ə̃}n	1
tɛ̃∧n	5	tɛ∧·ɪ>n	1
tɛ̆∧n	4	tɛ∨·n	1
tɛ∨{ə}n	4	tɛ∨>·n	1
tɪ∨{ɨ}n	3	tɛən	1
tɛ>·n	3	tɛ∨n	1
tɛ{ə>}n	3	tɪ{ə}n	1
tɛ·{ə>}n	3	tɪ∨n	1
tɛ̃·n	3	tɛ∧<n	1
tɪ̃·{ə̃}n	2		

bears strongly on how infants learn to identify words out of the continuous stream of speech, by recognizing transitional probabilities between syllables (Saffran, Newport, and Aslin 1996). Additional research by Maye, Werker, and Gerken (2002) has suggested that "distributional learning," recognition of frequency information in a bimodal array of sounds rather than transitional probabilities, also plays a role in phonological acquisition. Statistical learning commonly uses nonsense words, but its researchers appear so far to be committed to phonemes as objects for learning and thus have not challenged the idea of phonemes as categories that infants acquire out of the continuous stream of speech. Still, the importance that statistical learning research has assigned to transitional probabilities and frequency addresses and questions the simplicity of Pierrehumbert's exemplar model. Transitions between syllables may be the focus of learning, not the phoneme per se, and the distributional pattern of speech sounds may affect recognition rather than just a focus on a phoneme target. Findings about statistical learning certainly support the importance of frequency distributions that emerge from complex systems, because they confirm the importance of frequency information, especially within a distributional pattern.

The exemplar model moves toward complex systems, but Pierrehumbert carries over aspects of traditional linguistics like the assumption of a phoneme system in the notion of categories or labels. Another carryover is the idea that categories are prototypical, that is, that there is a central tendency among the exemplars that could be computed as an average and displays a Gaussian distribution (147). Hintzman's original article about exemplars in psychology had referred to schemas, a system more adequate to the variation we have seen because it does not involved averaging or statistically normal distributions (Kretzschmar 2009: 220–225). Pierrehumbert's prototypes implicitly recall the single output we might expect from a "standard" model, with random variation around it. Pierrehumbert's proposal for historical sound change on the basis of the distribution of exemplars thus suggests that the curve moves over time. While Pierrehumbert's exemplar model makes an improvement in the management of frequency information over Bybee's earlier high-level generalizations, it does not yet get all the way to complex systems because it addresses neither the nonlinear distribution of realizations nor their scale-free distribution.

Adele Goldberg (2006)

The next stop on the tour is Goldberg's 2006 *Constructions at Work*, which advances from her 1995 book, *Constructions: A Construction Grammar Approach to Argument Structure*. While it brings to bear another decade of work in the field, her 2006 book maintains the key point of her earlier work, that "all levels of grammatical analysis involve constructions: learned pairings of form with semantic or discourse function, including morphemes or words, idioms, partially lexically filled and fully general phrasal patterns" (Goldberg 2006: 5). Her other essential point is that "Constructions are combined freely to form actual expressions as long as they are not in conflict" (2006: 10). This of course differs sharply from formal models by rejecting linguistic nativism and biologically coded universals, and also by favoring an additive, unitized grammar of constructions as opposed to a system of rules which derives surface structure from underlying levels in the rule system. Still, as we saw in Bybee's (2001) work, Goldberg does value continuity with formal linguistics even as she attacks it on many fronts, as shown here (2006: 4):

> Constructionist approaches share certain foundational ideas with the mainstream "generative" approach that has held sway for the past several decades (Chomsky 1957, 1965,

1981). Both approaches agree that it is essential to consider language as a cognitive (mental) system; both approaches acknowledge that there must be a way to combine structures to create novel utterances, and both approaches recognize that a non-trivial theory of language learning is needed.

Language is still a system instantiated in the mind. On the other hand, Goldberg's treatment of frequency effects does advance beyond Bybee's. In a section called "Skewed input" Goldberg reports on a corpus of transcribed mother–child conversations (2006: 75–76):

> In analyzing the mothers' speech we found a strong tendency for there to be one verb occurring with very high frequency in comparison to other verbs used in each of the constructions analyzed . . . For example, *go* accounts for a full 39 per cent of the uses of the intransitive motion construction . . . This high percentage is remarkable since this construction is used with a total of thirty-nine different verbs in the mothers' speech in the corpus.

Besides *go* in the intransitive motion construction, Goldberg also cites *put* as occurring 38 percent of the time out of the forty-three verbs used in the transitive construction with motion, and cites *give* as occurring 20 percent of the time out of the thirteen verbs used in the ditransitive construction. This report of course matches what we should expect of nonlinear distributions in a complex system, although Goldberg does not make the connection to complex systems. Still, the combination of her specific reference corpus with recognition of the skewed input for the children's language learning allows Goldberg to make a general statement that, for the most part, is not tied to traditional formalisms (2006: 92):

> One particular verb accounts for the lion's share of tokens of each argument frame considered in an extensive corpus study . . . The dominance of a single verb in the construction facilitates the association of the meaning of the verb in the construction with the construction itself, allowing learners to get a "fix" on the construction's meaning. . . . In this way, grammatical constructions may arise developmentally as generalizations over lexical items in particular patterns.

Only the last sentence invokes the assumption of traditional linguistic practice, by locating "grammatical constructions" as the outcome of making "generalizations over lexical items in particular patterns." Otherwise, the statement captures a genuine and important distributional effect that arises from a particular group of speakers, and connects it to individual cognitive development.

The association of constructions with cognitive processing connects construction grammar with exemplar theory (e.g. Bod 1998; Skousen 1989). The manner of categorization for constructions also parallels exemplar theory in that categorization, now of grammatical forms in addition to speech sounds, avoids the reification of traditional categories, and is thus much freer than in earlier models. Both of these characteristics are important for complex systems, just not sufficient to bring Goldberg entirely into complexity science.

Michael Tomasello (2003)

Michael Tomasello's 2003 *Constructing a Language* was published three years before Goldberg, but in many ways Tomasello advances further toward a more usage-oriented model. However, he begins not far from the previous books on our tour (2003: 5): "in contrast to generative grammar and other formal approaches, in usage-based approaches the grammatical dimension of language is a product of a set of historical and ontogenetic processes referred to collectively as *grammaticalization*." Again, grammaticalization is not here what it was for Hopper in 1987, a process that never creates a product, but instead leads to "the grammatical dimension of language" which consists of patterns of use that "become consolidated into grammatical constructions" like the passive or noun phrases. Tomasello recasts grammar as an inventory of constructions (2003: 5–6): "A plausible way to think of mature linguistic competence, then, is as a structured inventory of constructions, some of which are similar to many others and so reside in a more core-like center, and others of which connect to very few other constructions (and in different ways) and so reside more towards the periphery." Gone here is the assumption that "the properties that distinguish sentences from vocabulary" should constitute the basic model for grammar for Hopper and Traugott. Gone too is the abstract structure of relationships that Bybee accepted from the generative model. These are replaced with a wider view in which the present inventory of an individual can be located with respect to "historical and ontogenetic processes."

Whereas previous authors on the tour address usage and frequency mostly at the top level of language in their discussions, Tomasello makes usage in particular learning environments into a very concrete factor for language acquisition. His skill of "intention reading" (sharing attention with others, following attention and gesturing of others, ability to learn the intentional actions of others) is an apt characterization of what children learn and what speakers do all the time: it describes the "joint attentional frame" that speakers must have in order for feedback to occur, which over time allows for the emergence of speech patterns both cognitively for individuals and socially in groups of speakers. Tomasello gives recurrence a central role (2003: 88): "If a child were born into a world in which the same event never recurred, the same object never appeared twice, and the adult never used the same language in the same context, it is difficult to see how that child – no matter her cognitive capabilities – could acquire a natural language." Recurrent situations of use – objects, events and context – are here explicitly joined to the "same language." The frequency of recurrence thus comes to the fore but, as for Pierrehumbert and Goldberg, not complex systems, which are not mentioned in the book. Tomasello suggests that frequency is not a monolithic factor in acquisition (2003: 79): "it is not the case that the more often a child hears a particular word, the quicker she learns it. As just two examples, the words *the* and *a* are among the most frequent words that young children hear, but they do not learn these words particularly early." This observation crucially invokes the importance of the process of language acquisition, not just frequency, in cognitive development.

It is well worth expanding on this point. Tomasello makes the general statement that "across many of the world's languages children initially seem to learn adult nouns more readily than adult verbs and other types of words" (2003: 45). The usual explanation for early learning of nouns and names is that these units are the easiest to associate with concrete aspects of experience (2003: 46–50). As the vocabulary grows, social terms greatly decrease while nouns remain as a large percentage. However, by age 2 token frequency begins to assert itself; in Tomasello's example list of the twenty-five most frequent words for one 2-year-old (2003: 82), there are no nouns and only one name – "Mummy." Tomasello explains that (2003: 81–82):

> these 25 words, which represent less than 2 percent of this child's vocabulary, account for 45 percent of all the words she says in terms of token frequency. What this list and these facts demonstrate is that as soon as just a small amount of grammar begins, young children's utterances are peppered with a relatively small number of high-frequency lexical items such as certain pronouns and function words with highly recurrent discourse functions – with the more well-known nouns and verbs, which are typically thought of as the prototypical items in young children's vocabularies, used relatively infrequently as their specific referents occur in the child's experience at only irregular intervals.

Thus learning begins with isolates (nouns, names), and then nonlinear frequency takes over. "Highly recurrent discourse functions" account for the distributional pattern according to Tomasello. *A* and *the* are on the list now for the 2-year-old because they recur with a great many nouns, while the situation for the use of any particular noun will recur only seldom. To see the nonlinear pattern indicates not merely the fact that the child has "just a small amount of grammar," but that the child has now passed the isolate stage and has achieved the threshold for normal linguistic interaction. These facts show that grammar is developmental, and of course the child's command of her language will continue to develop. Indeed, as Tomasello later observes about the acquisition of nominal constructions (2003: 198): "Children learn and use nominals in their very earliest holophrastic language, but true skill with nominals comes very slowly because successfully identifying referents for different listeners in different discourse contexts requires sophisticated skills of perspective-taking (theory of mind)." A good example of this point is Hopper's high- and low-frequency uses of the determiners *a* and *the*, which would take learners some

Table 3.2 *Mothers' and children's' construction frequencies by type (adapted from Tomasello 2003: 112, 148)*

Construction types mothers use with their 2-year-olds		
	Current study	Wells study
Fragments	.20	.27
Questions	.32	.22
Imperatives	.09	.14
Copulas	.15	.15
Subject–predicate	.18	.18
Complex	.06	.05
Adult categorization of utterances that 2-year-old children use with their mothers		
Fragments	.69	
Questions	.04	
Imperatives	.03	
Copulas	.07	
Subject–predicate	.16	
Complex	.002	

time to experience and acquire. Tomasello's phrase "true skill" is a good one, in that it does not mean that children acquire any fixed grammar of nominals, but instead implies that the developmental process of skill with nominals can continue for a lifetime.

The interaction of recurrent situations for language use, frequency effects, and children's developmental process of language acquisition are again illustrated nicely by Tomasello with charts of the relative frequency of construction types by mothers and by children (2003: 112, 148; see Table 3.2). On the children's' side, we see that the frequency data does not match what their mothers say. The children have many more fragments and fewer questions, predictably almost no complex constructions yet, though perhaps surprisingly about the same percentage of subject–predicate constructions as their mothers. On the mothers' side two experiments are represented, and the frequencies of constructions are not the same. The differences can be quite large, such as the Wells group with a third more fragments, a third fewer questions, and half again as many imperatives. The finding is just what a complex systems view would predict, that different groups of speakers will have different distributional patterns for the variants of linguistic features we track (here, constructions). The consequence of the finding for children acquiring language is, as Tomasello points out (2003: 110): "Children learn what they hear, and different children hear different things in different quantities." Tomasello uses this data to argue that "language acquisition is not just triggered by the linguistic environment, as proposed by generative grammarians, but rather the linguistic environment provides the raw materials out of which young children construct their linguistic inventories." Recurrence, frequency, and cognitive development may be hard to tease apart, but they all certainly appear in the data.

Nick Ellis and Diane Larsen-Freeman (2009)

The next-to-last stop on our tour is Nick Ellis and Diane Larsen-Freeman's 2009a *Language as a Complex Adaptive System.* The book begins with a position paper by the members of the "Five Graces Group," which includes Joan Bybee, William Croft (another major figure in usage-based linguistics, some of whose interests are not far from Goldberg's as in Croft 2001), and several others as well as the volume editors. With this roster of participants, complex systems has thus now become a focal mode for usage-based linguistics (Ellis and Larsen-Freeman 2009a: 5):

> A usage-based theory of grammar in which the cognitive organization of language is based directly on experience with language. Rather than being an abstract set of rules or structures that are only indirectly related to experience with language, we see grammar as a network built up from the categorized instances of language use . . . The basic units of grammar are constructions, which are direct form-meaning pairings that range from the very specific (words or idioms) to the more general (passive construction, ditransitive construction), and from very small units (words with affixes, *walked*) to clause-level or even discourse-level units . . . Because grammar is based on usage, it contains many details of co-occurrence as well as a record of the probabilities of occurrence and co-occurrence. The evidence for the impact of usage on cognitive organization includes the fact that language users are aware of specific instances of constructions that are conventionalized and the multiple ways in which frequency of use has an impact on structure.

The Five Graces clearly rely on terms and ideas that we have seen in usage-based linguistics. Constructions are the basic unit, the components in the complex system. Their "record of the probabilities" recalls exemplar theory. The Five Graces' assertion of a "network built up from categorized instances of language use" transforms Tomasello's (2003) "inventory of constructions" with a term from complex systems, "network." They also apply grammaticalization (6), and specify the goal of their position (7): "In the usage-based framework, we are interested in emergent generalizations across languages, specific patterns of use as contributors to change and as indicators of linguistic representations, and the cognitive underpinnings of language processing and change." Grammaticalization, then, for the Five Graces is the universal process that describes the operation of the complex adaptive system of speech. Their general statement on complex adaptive systems should also be familiar from the earlier discussion of complex systems:

> Language as a [complex adaptive system] of dynamic usage and its experience involves the following key features: (a) The system consists of multiple agents (the speakers in the speech community) interacting with one another. (b) The system is adaptive; that is, speakers' behavior is based on their past interactions, and current and past interactions together feed forward into future behavior. (c) A speaker's behavior is the consequence of competing factors ranging from perceptual mechanics to social motivations. (d) The structures of language emerge from interrelated patterns of experience, social interaction, and cognitive processes. (Ellis and Larsen-Freeman 2009a: 2)

Complex systems, then, appear to be a perfect match for usage-based linguistics.

Unfortunately, their final statement that the "structures of language" are the product of emergence from the operation of the complex system, carries forward the problems I have noted for the previous usage-based works, in that the Five Graces prejudge the outcome that may emerge from the complex system. In focusing on grammar as a kind of object, a "network," they convert the process that Hopper emphasized in "Emergent Grammar" into a state. For any single linguistic feature, the Graces refer to the process of "selection" as the emergence of a preference for one state of the feature, one variant, over other possible variants. They associate change with long-term alteration in social practices, "which in the extreme case leads to the fixation of [new forms] and extinction of [old forms]": "Changes in lifestyles lead to the rise and fall of words and constructions associated with those lifestyles (e.g., the rise of *cell* [phone] and the fall of *harquebus*). In the latter case, the social identity and the social contexts of interaction lead to the rise and fall of linguistic forms that are associated with various social values by speakers" (Ellis and Larsen-Freeman 2009a: 9). The Five Graces' preference for, or "fixation" of, a grammatical state takes the dynamic movement of the complex system and freezes it, so that one variant of a feature becomes "grammatical" in the sense of having been selected. The Five Graces Group, in my view, has been too eager to identify grammar directly with one aspect of complexity science, states and state space. And, as I have shown throughout the tour of usage-based linguistics, this problem arises because, as Walker Percy might have said, the authors presented here have stayed too close to the Scholastics and have not acted enough like Galileo in their observations. The position paper does move in the right direction, but it still carries some old baggage.

Complex systems is where usage-based linguistics ought to be going. We see in Figure 3.3, for instance, adapted from Ellis and Larsen-Freeman 2009b, good examples of the nonlinear distributional pattern in verb–argument constructions drawn from the speech of interviewers and learners in the European Science Foundation ESL corpus. For the verb–locative construction (VL), *go* is the most common verb by far in these constructions, with abruptly diminishing frequencies for other verbs, an A-curve. Only the top fifteen verbs are shown for both groups of speakers, so one can imagine a longer tail of uncommon

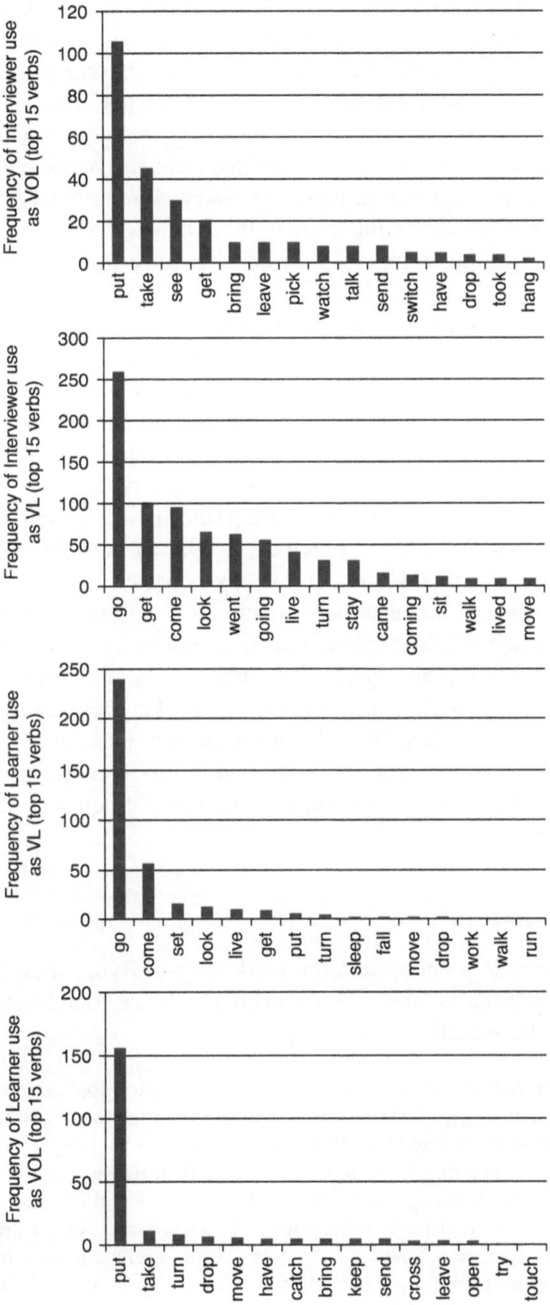

Figure 3.3 VL and VOL construction frequencies (adapted from Ellis and Larsen-Freeman 2009b:101)

verbs. Ellis and Larsen-Freeman invite us to "note the similar rankings of verbs across interviewers and learners" (2009b: 101) – but it is not exactly the same. Yes, *go* is at the top in both cases, and *come*, *look*, *get* are also common, but only nine of the fifteen verbs appear in both lists, and they are not in the same order. The same thing happens for the verb–object–locative constructions (VOL). *Put* is by far the most common verb for both groups, followed by *take*. However, this time only seven verbs appear on both lists, and again the order is different after *put* and *take*. What is truly similar about these lists is the A-curve pattern, with its evidence of scaling in subsamples, not the entire state of the network of nodes.

The discovery of the nonlinear pattern for the ESL teachers and learners in the ESF corpus is an important extension of what we know. The existence of nonlinear distributions does not prove that any category we have selected is real or valid, however, because the scaling property of complex systems tells us that we can expect to see a nonlinear distribution anywhere we look, at any level, for any phenomenon, for any group of speakers. To say, as the Five Graces do, that "grammatical categories develop [through grammaticalization] in all languages" (Ellis and Larsen-Freeman 2009a: 8) puts the cart before the horse: frequency distributions occur for any constructions we decide to nominate, but linguists are the ones who create the categories, who make the grammar. The operation of complex systems does not create a network out of which we observe a state as an object. The complex system merely creates a nonlinear frequency distribution, and linguists think that they see a grammar in the distribution after the fact. Ellis, Larsen-Freeman, and the Five Graces, for all their movement in the right direction, still fail to distinguish their own perceptions of structure from the continual emergence of frequency patterns in the complex system.

Joan Bybee (2010)

Our last stop on the tour is once again the work of Joan Bybee, now her 2010 book *Language, Usage and Cognition.* In this book Bybee frankly champions language as a complex adaptive system (2):

> When linguistic structure is viewed as emergent from the repeated application of underlying processes, rather than given a priori or by design, then language can be seen as a complex adaptive system . . . The primary reason for viewing language as a complex adaptive system, that is, as being more like sand dunes than like a planned structure, such as a building, is that language exhibits a great deal of variation and gradience. Gradience refers to the fact that many categories of language or grammar are difficult to distinguish, usually because change occurs over time in a gradual way, moving an element along a continuum from one category to another . . . Variation refers to the fact

that the units and structures of language exhibit variation in synchronic use, usually along the continuous paths of change that create gradience.

The idea of emergence, as from Paul Hopper's 1987 paper, has now returned to become a centerpiece of the usage-based approach. The two other key terms here, gradience and variation, refer to adaptation of categories over time, and to the simultaneous presence of alternative usages at any given time. The notion of a continuum of change confirms (but does not cite) what we saw in Chapter 1 from Burkette (2001, 2009).

Bybee (2010: 19–22) has now adopted the exemplar model for phonology that we have seen from Pierrehumbert. She has also embraced constructions (2010: 25–32), which she also describes according to the exemplar model. Significantly, she says that (2010: 25):

> Constructions also have exemplar representations, but these will be more complex, because, depending upon how one defines them, most or all constructions are partially schematic – that is, they have positions that can be filled by a variety of words or phrases.

The reference to schemas moves beyond her earlier discussion of prototypes (2010: 18–19), which Pierrehumbert had also treated, as a model for categorization in language. Her discussion of how exemplars of constructions arise (2010: 28–31) maintains a nuanced view, from fixed idioms to the connotations that a turn of phrase like "what are you doing?" may have (in this case negative ones), to the emergence of variations on a turn of phrase like "what's a nice girl like you doing in a place like this?" As she later points out, the cognitive associations of the emergence of constructions "underscore the importance of multi-word chunks" (2010: 38), both to phonology and grammar, as a way that exemplars function. Bybee's description of "fixedness vs. schematicity in the slots of constructions" (2010: 80–81), in conjunction with "prefabs" (2010: 81–83), exemplifies what she means by gradience and variation. These are all strong, positive moves toward complex adaptive systems.

At the same time, however, Bybee still talks about prototypes (2010: 83–88). She discusses grammaticalization (2010: 106–114) as the process "by which grammatical items and structures are created" (2010: 106), a statement which again backs away from continual emergence. Later chapters discuss traditional categories like the English auxiliary and Modern English *can*, larger, top-level generalizations of the kind from Bybee (2001), which take Bybee away from the local and schematic properties of constructions discussed in earlier chapters. In her final chapter, one of the sections talks about "grammaticalization paths as 'strange attractors'" (2001: 198), which does not apply the technical term "strange attractor" from complexity science in the way it is usually used for chaotic patterns. By the end of the book, while Bybee confirms her interest in "domain-general processes" that transcend individual languages, the operation of the complex system of speech may for some readers have become less clear rather than precisely illustrated. Bybee does not yet deal extensively with scaling or with characteristic frequency profiles. She has become the most advanced advocate of complex adaptive systems among usage-based linguists along with Ellis and Larsen-Freeman, and yet her work still has a way to go to escape from the Scholasticism of linguistics that Walker Percy raised, and to become more like the Galileo that Percy said we needed. As with the Five Graces, Bybee continues to carry old baggage from formal linguistics.

Usage-based linguistics and complex systems

So, at the end of the tour, what have we learned about usage-based linguistics and complex systems? First, all of the evidence from Chapter 1 indicates that we can expect nonlinear distributions of variants for any linguistic feature we are interested to follow. This is true of speech sounds, whether measured impressionistically or acoustically, of words, and of constructions, as additional evidence presented in this chapter has shown. There was an early champion of emergence in Paul Hopper's 1987 paper, but that early position was not well maintained as usage-based linguistics developed. As Hopper and Traugott (1993) and then Bybee (2001) have demonstrated, linguistic categories are not naturally given and discrete: over time expressions change in such a way that any boundaries for categories that we may wish to assert at one moment may well be breached a moment later. Exemplar theory suggests a cognitive process by which we can build frequency profiles for speech sounds, although the process of categorization remains an issue there. Still, we need to use categories so that we can count variant expressions and observe the nonlinear distributions. The idea of "constructions" is perfect for this purpose, since the use of constructions does not entail a commitment to a fixed hierarchy of categories, and it may be applied to features as small as speech sounds or as large as discourse patterns. There is no reason that we cannot use traditional terms like "noun" or "verb" or "ditransitive" to name them, as long as we do not invoke an entire hierarchical system when we do so. Grammaticalization really is just what Hopper said it was in 1987: a process of continual movement that is contingent on time, place, and circumstance and that does not allow grammar ever to be directly observable; it does not translate into categories for any state of any language. Neither does the nonlinear distribution constitute evidence that there is a particular cause for the top-ranked variant to be where it is. The complex interaction of recurrence, frequency, and setting for language use rules out any simple cause for the state of a feature or language at any given time, except to say that the process of the complex system of speech always creates nonlinear distributions.

Equally important, all of the evidence shows us that scaling is an unavoidable property of the complex system of speech, whether from the Linguistic Atlas, Tomasello (2003), or Ellis and Larsen-Freeman (2009b). The property of scaling has the unfortunate effect that we will find it difficult to generalize from one level of the language to any other. It is just not credible to jump back and forth between frequency evidence on a national level, like the Francis and Kučera rankings, and cognitive development in individuals. The operation of the complex system of speech in the cognitive development of children is the same in principle but essentially different in outcome from the operation of the complex system in a speech community at any level. As Tomasello has said, "children learn what they hear, and different children hear different things in different quantities" (2003: 110). We can never assume that others will share our own experiences with constructions. If we take care to match the assessments we make to the particular populations from which our data comes, we can make better generalizations, whether for a language as a whole, for national or regional varieties, for social groups, or for particular kinds of texts. The complex interaction of recurrence, frequency, and setting for language look different from every point of view at every scale of analysis.

Finally, speakers as agents using speech are not dull bodies who have no choice in what features they want to use on any given occasion, including all of the different kinds of conversations and text types that people enact, and so the complexity of speech is much, much greater than we have so far given it credit. The very idea that any single grammar could describe or generate such complexity is, frankly, beyond imagination; a complete single grammar for a language could never be motivated by detailed observation of speech production. And of course the "standard" generative model and other formal models are not so motivated, and indeed were created to avoid the complications of language in use and to emphasize what might be the same about speech. Formal linguists have good reason to want to generalize away from observations of language production, and we can make good use of reasonably constructed abstract grammars, as discussed in the next chapter.

Usage-based linguists also have good reason to want to observe language in use. When we are aware of the magnitude of the distributional problem, we certainly know that whatever experiment we conduct will never be enough to give us the kind of big-picture answers that formal linguists have generated. Usage-based linguists should stop trying to align with formal linguists because formalists are trying to answer questions that are essentially different from those based in the observation of language in use. We may discover some cross-linguistic similarities, as usage-based linguists have already claimed to have done. Indeed, the logical conclusion we can draw from complex systems is that there is really only one language, the human language, and the phenomena that we have perceived as different languages are actually levels of scale within the overall complex system of human language. To do a good job of language typology or cross-linguistic studies, we cannot ignore the complexity that we know to exist in aggregated data sets at any level of scale. Nobody can do it all alone. Walker Percy really needed more than just one Galileo to observe all the moons, planets, and stars in the linguistic cosmos.

Together, linguists who observe language in use, whether in settings that feature acquisition or adult speech, can build up a better understanding of the complex system of language. If we stop trying to bring along the old baggage of fixed grammars and selected usages, we are free to choose constructions to count and free to see nonlinear frequency patterns wherever they occur. Every experiment that we conduct, if it adequately describes the constructions it studies and the population of speakers who use them, makes another contribution to our knowledge of the complex system of language in use. The more evidence from well-controlled studies that we can build up, whether from small populations or from the sampling of large groups, the better picture we can develop of the way people use language at every scale of interaction. Industrial applications such as human–computer interfaces will certainly benefit from a knowledge of scale-free effects, as opposed to the general application of a top-level grammar. What we stand to gain more generally by such repeated studies, even though we know in advance that studies at different scales of analysis will not be comparable in their findings, is a new understanding of the operation of complex systems in human culture. The big picture is not a grammar, but instead a new way of understanding the humanities as the emergence of linguistic and cultural patterns out of continual human interaction. Continual movement at every level of scale is crucial to that understanding. Paul Hopper launched usage-based linguistics in reference to cultural anthropology, and continuing observations by usage-based linguistics can confirm the correctness of his insight.

4 Grammar and complex systems

The usage-based linguist Michael Tomasello has compared a number of theories about language acquisition (2003) including what Pierrehumbert in the last chapter called the "standard" model, generative linguistics. Tomasello is not an admirer of the work of Chomsky and Pinker. He begins with the following characterization of the difference between "formal linguistics" and usage-based linguistics (98–99):

> formal linguistic approaches (including generative grammar) characterize natural languages in terms of formal languages, using as basic theoretical primitives meaningless algebraic rules and meaningful linguistic elements that serve as variables in the rules. But many contemporary linguists simply do not believe that the analogy between natural languages and formal languages is a particularly accurate or productive one – most importantly because it effaces the symbolic dimension of grammatical constructions. The alternative is to look at linguistic competence, not in terms of the possession of a formal grammar of semantically empty rules, but rather in terms of the mastery of structured inventory of meaningful linguistic constructions.

The previous chapter has provided the basic terms for the alternative approach, including exemplar theory and constructions. Tomasello sums up the difference between the "standard" model and its alternative (2003: 100):

> The most important point is that constructions are nothing more or less than patterns of usage, which may therefore become relatively abstract if these patterns include many different kinds of specific linguistic symbols. But never are they empty rules devoid of semantic content of communicative function. In usage-based approaches, countless rules, principles, parameters, constraints, features, and so forth are the formal devices of professional linguists; they simply do not exist in the minds of speakers of a natural language.

Patterns of usage, then, not formal rules, constitute the grammar of a language for Tomasello as it inheres in the human cognitive processes. Still, the patterns may "become relatively abstract" under the right conditions, moving towards the algebraic formulas of the generativists, and indeed, generativists would insist that all linguists have an empirical side, so perhaps the difference between usage-based and formal models might consist in how the patterns are formulated. As discussed in the last chapter, usage-based linguists have not broken away from the old baggage of formal linguistics as much as they might like to think.

Tomasello does answer the generativist claim about the biological innateness of grammar (284):

> The best-known theory about the role of biology in human linguistic competence is Chomsky's proposal that with respect to core grammar biology is everything. In this theory, the innate language module (universal grammar) does not contain things like special learning procedures and perceptual biases but rather real linguistic content; it is thus a theory of "representational innateness" . . . But representational innateness is a very unlikely theory.

Tomasello then treats grammar genes, linguistic savants, brain localization, the critical period, deficient input, and poverty of the stimulus as reasons to believe in representational innateness, and finds all of them lacking. He concludes (289):

> Overall, then, the case for linguistic nativism – in the form of representational innateness – is very poor. Combining all the data and arguments throughout this book, we can say that: (1) there are virtually no linguistic items or structures that are universal in the world's languages; (2) there is no poverty of the stimulus in language acquisition; (3) linking does not work; (4) parameters do not help; (5) the continuity assumption is demonstrably false; (6) performance factors and the maturation of universal grammar are simply unprincipled fudge factors used to explain recalcitrant data; (7) invoking extensive lexical learning as necessary for triggering parameters makes the theory basically indistinguishable from other learning theories – except that it has in addition the linking problem; and (8) although the empirical situations cited in support of biological bases for language acquisition mostly do demonstrate such bases, they do not demonstrate in any form representational innateness.

This is quite an indictment, and yet as before it does not completely dispose of the generative model. We may even agree with Tomasello that the claim for biological innateness cannot stand, but we could still uphold the logical bases for the generative model. After all, most generativists spend little time arguing for representational innateness, and a great deal more time describing relationships within sentences of a language. The principle of "elegance" guides the logic of generativism, and elegance is a matter separate from representational innateness.

What I propose to do in this chapter is to suggest why we in language studies ought to pay more attention to what I will call the 80/20 Rule than we have before now, especially for the way that we understand the grammar of English. We can, and indeed will have to, accommodate more than one theory of grammar in the environment for language study that we have inherited. We should be prepared to cope with Tomasello's usage-based approach and related historical theories from structural grammar, and with traditional prescriptive school grammar, and with the "standard" generative model. Once we accept that there are in fact other different approaches to grammar that compete with generativism in our linguistic marketplace of ideas (see Kretzschmar 2009: 6–20), we need to discover the principles of each one so that we can assess them appropriately. Complex systems offers us the means to do these things.

Zipf's Law and the 80/20 Rule

The distributional pattern in language known as Zipf's Law has been around for nearly one hundred years now. Jean-Baptiste Estoup may have noticed the pattern first around 1912 (Petruszewycz 1973), but Zipf made it famous about three decades later in association with what he would come to call the "Principle of Least Effort" (1949). Wikipedia gives us the basic low-down (en.wikipedia.org/wiki/Zipf's_law):

> Zipf's Law states that given some corpus of natural language utterances, the frequency of any word is inversely proportional to its rank in the frequency table. Thus the most frequent word will occur approximately twice as often as the second most frequent word, which occurs twice as often as the fourth most frequent word, etc. For example, in the Brown Corpus "the" is the most frequently occurring word, and by itself accounts for nearly 7% of all word occurrences (69,971 out of slightly over 1 million). True to Zipf's Law, the second-place word "of" accounts for slightly over 3.5% of words (36,411 occurrences), followed by "and" (28,852).

In any text of at least moderate size, the frequency of any word in the text is just about inversely proportional to its rank. This statistical relationship is noticed in two of the best modern books on Natural Language Processing, a field which specializes in quantitative measurements of language, but the key word here is "noticed." Manning and Schütze (1999: 23–29) offer a concise, six-page account of Zipf's Law early in their very long book, using the text of Mark Twain's novel *Tom Sawyer* for illustration, but then they mention Zipf on only two occasions after the initial description, both in passing. Jurafsky and Martin's textbook *Natural Language Processing* (2000) does not even mention the Zipfian distribution until a passing reference on page 603, and that is the only mention of Zipf in the volume. Those of us in language studies who want to learn more about Zipf's Law will be faced with explanations like this one from PlanetMath (http://planetmath.org/zipfslaw):

> Zipf's Law says that the *i*th most frequent object will appear $1/i^{\theta}$ times the frequency of the most frequent object, or that the *i*th most frequent object from an object "vocabulary" of size V occurs
>
> $$\mathrm{O}(i) = n / i\theta H_{\theta}(\mathrm{v})$$
>
> times in a collection of *n* objects, where $H_{\theta}(\mathrm{V})$ is the harmonic number of order θ of V.

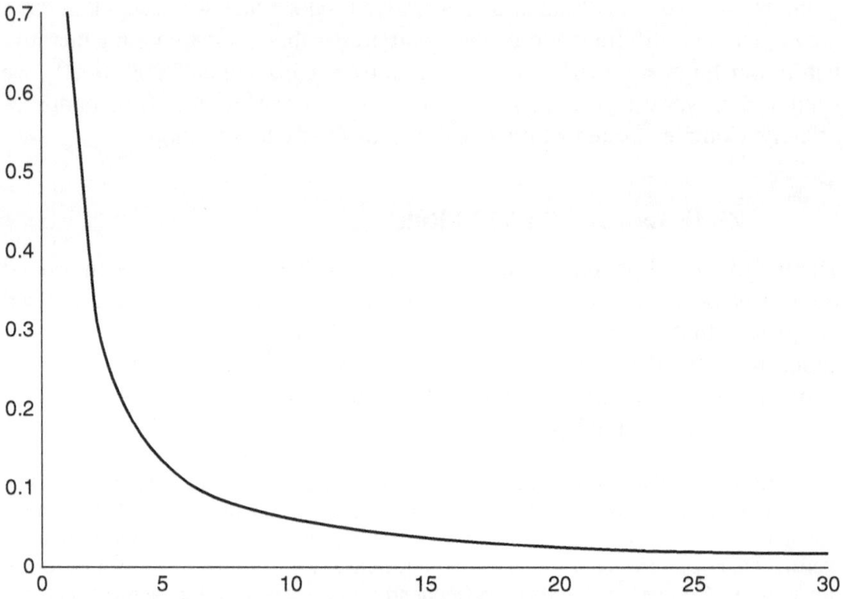

Figure 4.1 Zipf's Law (adapted from PlanetMath, http://planetmath.org/zipfslaw)

PlanetMath is a website that bills itself as "math for the people, by the people," but whose language and notation may leave many non-mathematicians behind. Figure 4.2, however, is a bit more intelligible, especially if we call it an "A-curve" as described in Chapter 1. If we put all the different word types in a text or corpus in order by frequency, there will only be a few word types that occur very often, at the left of the graph, and there will be a great many word types that do not occur very often at all, at the right of the graph in what is often called its "long tail." Let us say that each of the thirty gradations on the PlanetMath curve represents one hundred different words. Then, as in Figure 4.3, the first five or six gradations at the left of the graph, about 500 or 600 words, account for about 80 percent of all the running words in the text, while the remaining 24 or 25 gradations in the long tail at the right, the other 2,400 or 2,500 different words, account for only about 20 percent of the running words. This is the source of the so-called "80/20 Rule," known as the Pareto Principle in economics from Pareto's early-twentieth-century observation that 80 percent of the land in Italy was owned by 20 percent of the population. The 80/20 Rule is often used in business (e.g. 80 percent of profits come from 20 percent of customers), quality control (e.g., fixing the top 20 percent of mistakes will control 80 percent of the complaints), and elsewhere. We in language studies can and should make good

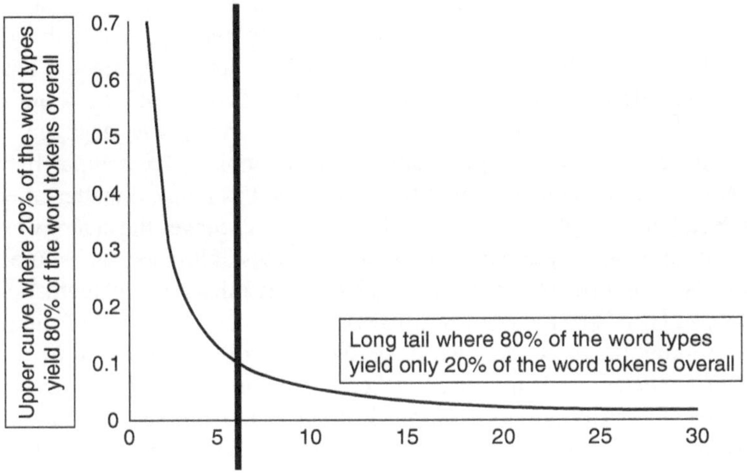

Figure 4.2 The 80/20 Rule

practical use of the 80/20 Rule on a conceptual basis, with the understanding that there is real mathematics behind it that derives from the complex system of speech.

In this practical vein, perhaps the most entertaining account of Zipf's Law was written by Hugh Kenner, in an essay called "Neatness Doesn't Count After All – Tidy vs. Untidy Desks" (1986). Kenner justified his practice of keeping an untidy desk with Zipf's Principle of Least Effort:

> untidy-deskers of the world may take heart. It's we who have mathematical validation. Forget what you may have thought about the swept and tidy world of numbers. Concentrate on the fine randomness of Einstein's hair. We connoisseurs of scrutable chaos have been guided all along by an inscrutable proposition called the 80–20 rule: a special case of Zipf's Law.

That is to say, if we keep the 20 percent of our books that we want 80 percent of the time near us on our desks, we save effort, as Kenner says, because they are "instantly within reach from where you sit. At the cost of a little rummaging, of course." The more serious discussion in the Kenner article mentions the use of the 80/20 Rule by IBM to optimize data processing in information science, and he offers examples of Zipf's Law in practice for literary works from Henry James, T. S. Eliot, James Joyce, and Shakespeare. In the end, Kenner suggests that "Zipf and 80–20 end up sketching what actually goes on, quite as if they were laws of impersonal Nature, like gravity and thermodynamics. But at bottom – so Zipf assured us – such laws work because they describe the situation we create in the course of intelligent coping. Situations like my desktop." While there are many reasons why you might like to keep particular books on your desk, Kenner's wit found a practical function for the Zipfian distribution there – at the cost of a little rummaging. Whether we use mathematical formulae or Kenner's desk as our mental model, we can see that Zipf's Law and the 80/20 Rule describe a distributional pattern that we can apply to language.

The problem with Zipf's Law is that it implies that regular behavior, in nature and in language, ought to be described as a law, a matter of cause and effect, here the Principle of Least Effort. Kenner makes the analogy explicit by mentioning gravity and thermodynamics, "laws of impersonal Nature." In fact, Zipf's formula has been superseded by, among others, the mathematician Mandelbrot (who spent his career at IBM). Mandelbrot's improved formula (1968) shows that the top-ranked words on the curve deviate from the frequency that Zipf expected, and the lower-ranked words also deviate, owing to what he called "the wealth of vocabulary." In Linguistic Atlas survey data (not written words in continuous discourse but spoken words and phrases gathered in the field), the top-ranked variant is often three, four, five, even ten times more frequent than the second-ranked variant, and we also see curves that are shallower than a 2:1 ratio between the first and second variant (discussed in depth in Chapter 7). If Zipf's Law were really a law, in the same way that thermodynamics and gravity are natural laws, then Zipf's Law just does not work well enough. Frequency is not always and exactly inversely proportional to rank. Hugh Kenner said as much in his article, when he reported that the 80/20 Rule in Henry James was actually 73/27, and in T. S. Eliot it was actually 71/35. What is really important here, however, is not the Principle of Least Effort or any formula of inverse proportions, but the shape of the curve, whether the proportion is actually 80/20 or 90/10 or 73/27. We know that a relatively few types will account for most of the tokens (which we might call "the 80 percent group"), and the remaining types will account for a much smaller quantity of the tokens (which we might call "the 20 percent group"), given some margin for variation in the actual ratio.

English grammar

Let us now turn to English studies and to grammar. If we begin with the long view, Saussure has described the history of "grammar" (1916/1986: 1):

> This discipline, first instituted by the Greeks and continued mainly by the French, is based on logic. It offers no scientific or objective approach to a language as such. Grammar aims solely at providing rules which distinguish between correct and incorrect forms. It is a prescriptive discipline, far removed from any concern with impartial observation, and its outlook is inevitably a narrow one.

"Traditional grammar," as Saussure later calls it, is "concerned with the description of linguistic states," and not at all with any historical development in a language. It fails to be scientific because (1916/1986: 82)

> Traditional grammar pays no attention to whole areas of linguistic structure, such as word formation. It is normative grammar, concerned with laying down rules instead of observing facts. It makes no attempt at syntheses. Often, it even fails to distinguish between the written word and the spoken word.

The implication here is that scientific observation is systematic and comprehensive, and traditional grammarians leave too much out of account in their attempt selectively to lay down rules. Nonetheless, we must all cope with traditional grammar because it is embedded in our educational systems, not just for the French but everywhere. Traditional grammar, then, is not a product of the complex system of speech, but instead arises from the unscientific opinions of grammarians.

Our popular belief in the prescriptive approach to grammar has generated a parallel belief in "standard" forms of languages. Besides prescriptive teaching in the schools, language standards are also recognized in the law courts. So, in the opinion of the judge in the famous Ann Arbor decision, C. W. Joiner equated the school "standard" with the language of "the commercial world, the arts, science, and professions" (*Martin Luther King Junior Elementary School Children et al.* v. *Ann Arbor School District*, 73 F. Supp. 1371 (ED Mich. 1979); italics added):

> The problem in this case revolves around the ability of the school system, King School in particular, to teach the reading of *standard English* to children who, it is alleged, speak "black English" as a matter of course at home and in their home community (the Green Road Housing Development). This case is not an effort on the part of the plaintiffs to require that they be taught "black English" or that their instruction throughout their schooling be in "black English" or that a dual language program be provided... It is a straightforward effort to require the court to intervene on the children's behalf to require the defendant School District Board to take appropriate action to teach them to read in the *standard English* of the school, the commercial world, the arts, science, and professions.

Judge Joiner distinguished between "black English" and "standard English," and clearly thought that "black English" deviated from the "standard English" that he believed to be in use in school and in educated realms of society. For Judge Joiner, "standard English" constituted the normal language of his own community of educated people, to which students in the school should aspire even if their own (deviant) home and community language differed from it. US Supreme Court justice Antonin Scalia has similarly advocated the "plain language rule" for the interpretation of statutes and of the constitution.

The idea that there is such a thing as a "plain meaning" to be drawn from a statute requires the implicit assumption that the language is somehow standard enough to support such a meaning for all readers. The language of the law has historically been notorious for convolutions designed to make sure that only one meaning can be derived from, say, a contract, which suggests that plain language may be more difficult to interpret than it is sometimes given credit. Still, the idea that plain language should have a plain meaning, as a result of our belief in linguistic standards, obtains still in the law and elsewhere in life around us.

The reference to "the Greeks" and to "logic" by Saussure is more complex. As John Dinneen has remarked (1967: 87):

> Grammatical distinctions . . . are not explicitly grammatical as opposed to logical in Aristotle's works, since he did not know of this distinction. The purpose he had in mind was logical, but the definitions he gives of the forms he uses in his logic, as well as the examples he uses, justify our considering these distinctions as grammatical.

Grammar and logic have remained linked for many teachers in Western educational institutions. As Dinneen remarks (1967: 71), "The terminology of traditional grammar, inherited from the Greeks, comprises the most widespread, best understood, and most generally applied grammatical distinctions in the world." Thus, as Saussure said, traditional grammar, from its origins among the Greeks, came to be linked with logic and to be characterized by prescription. It remains the primary mode of language instruction in our elementary and secondary schools, and as such we cannot ignore it.

Of course, modern linguists are not bound by the prescriptive tradition, and we do try to make observations scientifically, based on the evidence before us. Still, we have mainly preserved traditional grammatical distinctions and categories, from parts of speech to functional categories. We still talk about the subject and predicate, as Aristotle did, or about subject–verb–object structure, or about NPs and VPs. Moreover, we still think of these as a hierarchy of rules, things that we do in English (or another language) as opposed to things that we do not do. In their *Comprehensive Grammar of the English Language* (1985), Quirk, Greenbaum, Leech, and Svartvik wrote that (11–12)

> It is GRAMMAR that is our primary concern in this book. Words must be combined into larger units, and grammar encompasses the complex set of rules specifying such combination.

The "complex set of rules" in the *Comprehensive Grammar* turn out to be not complex enough: we maintain professional journals that continually publish additional claims for what Quirk et al. called "the regularities of grammar." Structural grammars like the *Comprehensive Grammar* thus must keep expanding in size, even though the *Comprehensive Grammar* already has 1,779 pages. The idea of a structural grammar thus includes documentation of as many regularities of grammar as might be identified in a language. Structural grammars must allow for growth in the grammar.

Academic grammars following Chomsky are not essentially different from this idea, but take grammatical complication in a different direction. Chomsky suggests that the central problem is deciding which generalizations are significant (1965: 41):

> We have a generalization when a set of rules about distinct items can be replaced by a single rule (or more generally, partially identical rules) about the whole set, or when it can be shown that a "natural class" of items undergoes a certain process or set of similar processes. Thus, choice of an evaluation measure constitutes a decision as to what are "similar processes" and "natural classes" – in short, what are significant generalizations.

Chomsky called generalizations that render the description of a system more complicated "spurious generalizations" as opposed to significant ones, which of course reversed the structuralist process of evaluation that has always led to bigger grammars, as structuralist grammars keep adding newly noticed grammatical constructions to the list. The distinction between significant and spurious generalizations underlies the generative principle of elegance. The generative trend in grammar seeks the fewest number of rules which can account for the regularities of, first, *a* language like English, on the way to universal rules that can account for human language in general (Chomsky 1965). In the end, however, Huddleston and Pullum 2002, a full-featured generative grammar that parallels the *Comprehensive Grammar*, is not smaller in size at 1,860 pages in the same large format. Generative grammars may have a principle of elegance that is supposed to control growth of the grammar by distinguishing significant generalizations from spurious generalizations, but in practice the principle of elegance has not led to smaller grammars.

What linguists have called a "complex set of rules" is just not the same thing as a complex system. Östen Dahl (2004) has written a book about the growth and maintenance of linguistic complexity, but that topic concerns the enrichment of possibilities with a hierarchy of rules, not complexity theory as described in this book. Indeed, the basic principles underlying grammar are just the opposite of those for a complex system:

1 grammar is a static structure of rules, as opposed to being open and dynamic;
2 grammar consists of a hierarchical arrangement of rules, ideally the smallest possible set of rules, as opposed to a very large number of interactive components/agents;
3 grammar exhibits fixed relations between elements, as opposed to emergent order. Because of these fixed relations, grammar ideally has binary

distributions, right vs. wrong, acceptable vs. unacceptable constructions, instead of the nonlinear frequency distributions characteristic of complex systems; and

4 grammar has homogeneous unity, as opposed to the property of scaling.

Grammar, then, as a way of thinking about language, is complementary to the complex system of speech, not a description of it. The properties of a grammar consist of idealizations, generalizations away from the dynamic qualities of speech as people actually use it. Generative grammar ideally should apply the principle of elegance to create the smallest number of rules that can generate the acceptable sentences of a language, while structural grammar collects regularities that need to be included in the static structure and thus allows for growth in the grammar. Traditional grammar allows grammarians to choose what rules should be included, without any necessary limitation by what people actually say or write in their language.

The "Five Graces Group" has proposed, on the basis of complex adaptive systems:

> a usage-based theory of grammar in which the cognitive organization of language is based directly on experience with language. Rather than being an abstract set of rules or structures that are only indirectly related to experience with language, we see grammar as a network built up from the categorized instances of language use . . . The basic units of grammar are constructions, which are direct form-meaning pairings that range from the very specific (words or idioms) to the more general (passive construction, ditransitive construction), and from very small units (words with affixes, *walked*) to clause-level or even discourse-level units.

The Five Graces Continue:

> Because grammar is based on usage, it contains many details of cooccurrence as well as a record of the probabilities of occurrence and co-occurrence. The evidence for the impact of usage on cognitive organization includes the fact that language users are aware of specific instances of constructions that are conventionalized and the multiple ways in which frequency of use has an impact on structure. (2009a: 5)

There are many things to agree with in this programmatic statement, but the direct connection of speech and grammar is not one of them. The Five Graces' assertion of a "network built up from categorized instances of language use" bears within it the notion of categorization, which implies some sort of a priori grammatical hierarchy, the "natural classes" in grammar cited earlier from Chomsky. The natural classes that Chomsky hopefully predicted to exist in 1965, linguistic universals, have not thereafter actually been documented (as Tomasello argued at the beginning of this chapter). Without natural classes, the process of categorization is fraught with problems that arise from all of the variation in speech around us, as we saw in the discussion of exemplar theory in Chapter 3. Indeed, in modern cognitive science, the theory of mental representations and the language of thought as championed by Jerry Fodor (see Bermudez 2010: 156–158, 251–253) is just one way of imagining the cognitive process, and categorization into representations is thus not something that we should accept as a given. The neural network theory of activation patterns (Bermudez 2010: 215–245) also explains how information and its frequencies can be processed, without categorization into representations. The newest work on the dynamical systems model of cognitive processing explicitly dispenses with representations (Bermudez 2010: 414–431; Van Gelder 1998 is a key source) in favor of the process involved in complex systems, interactions with feedback (see Chapter 6 for further discussion of cognitive processing). Thus the Five Graces Group is right to insist on usage as what builds a speaker's cognitive sense of a language, but not credible in their assertion of a direct connection between speech and grammar as a network of categories. Grammaticalization is a process without an endpoint, whether for speech in society as Hopper argued in 1987 or for cognitive information processing by individuals. Grammar, on the other hand, when it is defined as a network or hierarchy of categories or rules, is something essentially different from the output of the complex system of speech, something only indirectly related to language in use.

Our interest in making grammars that transform the complex system of speech into a hierarchy of rules comes originally, in our Western culture, from just the motivation that Saussure talked about. First, we have Aristotle's wish to specify how language can be used to make particular logical, philosophical relations between subjects and predicates. By extension, the logic of structure in a language tells us about what Saussure called "linguistic value" (see Kretzschmar 2009: 53–55). Linguistic value is the heart of Saussure's *Course in General Linguistics* (1916/1986), because linguistic value is the aspect of language that we can only describe through analysis of a fixed structure. Linguistic values are established as a result of the relational fit of each linguistic feature within the hierarchical system of the language. Thus, in Saussure's example (1916/1986: 114), the French word *mouton* covers the same semantic space as both English *sheep* ('live animal') and English *mutton* ('meat from the animal'), and so the French word is the equivalent of neither English word. The French system has one element while the English system has two, and the linguistic values must therefore be different because of the relational fit of the elements. The same sort of difference in linguistic values obtains for elements in both paradigmatic structures such as the list of words in a language, and also in syntagmatic structures such as the constructions into which those words may be assembled. Linguistic value can only arise from a hierarchical structure, whether a paradigmatic inventory of words or an array of syntagmatic rules. This means that it does not, cannot, come directly from a dynamic complex system with emergent order. Grammar is something that requires us to abstract away from what people are doing, to get outside of the process of interaction of the complex system, in order to build a hierarchy. Thus a grammar must by necessity rely on linguists' perceptions of what people are doing for generative and structural grammars, or rely on traditional grammarians' opinions of what people ought to be doing. Still, linguistic value will always justify an interest in linguistic structure as an appropriate mode for study of a language, even if we cannot extract it directly from the complex system of speech.

The real problem with grammars has always been the relation of a grammar to the complex system of speech, to language as we use it. Ideally, grammatical rules are binary: they tell us what we can do in a language and what we cannot do; we think that speakers of a language should be able to make grammaticality judgments. As Quirk et al. said of speech sounds in the *Comprehensive Grammar* (1985: 11),

> When people speak, they emit a stream of sounds. We hear the sounds not as indefinitely variable in acoustic quality (however much they may be so in actual physical fact). Rather, we hear them as corresponding to one of the very small set of sound units (in English /p/, /l/, /n/, /I/, /ð/, /s/ . . .) which can combine in certain ways and not others.

Thus the nature of rules, which rely on our perceptions. Anybody making a grammar will assemble a perceived set of categorizations from the stream of speech into a structure of combinatorial patterns that people use ("certain ways and not others," whether by observation or opinion). With regard to word-formation, Quirk et al. say that "products of word-formation are subject to the idiosyncrasies inherent in the lexicon, whereas the rules of grammar are far broader, transcending the individualities and unpredictabilities of the lexical items which the grammatical rules manipulate" (1985: 1518). Unfortunately, as speakers of English know very well, the "sound units," or phonemes, of English are not as clear-cut as this statement suggests. The inventory of phonemes is not the same for all speakers of English, and speakers of English also do not all share the same rules for combinations of sounds, say for the many speakers who vary in rhoticity, or who employ consonant cluster simplification, or who substitute glottals for other stops. The same is true of rules of grammar. Quirk et al. offer the "Rules for forming open-class *-ly* adverbs from adjectives," but also cite exceptions like *wholly* for expected *wholy*, *gaily* for expected *gayly*, *publicly* for expected *publically* (1985: 439). They talk about rules for clause formation, and then say that "These restrictions are frequently violated in poetry and in other imaginative uses of language" (1985: 772). In their discussion of pro-forms, with regard to constructions that they have earlier described "in terms of grammatical rule," they say that "What is grammatically tolerable has to be mediated by what is textually tolerable" (1985: 1460). So, even if grammatical and phonological rules are supposed to be binary, in fact they have never been so in structural grammars like the *Comprehensive Grammar*.

Newer generative approaches to rules do not allow exceptions or leaks, but instead adopt a strategy of rule ordering, or of multilevel surface and deep rules, or other ingenious manipulation of the hierarchy of rules in order to compensate for the basic fact that binary decision trees are a poor match for language in use.

Sapir has famously written that "All grammars leak" (1921: 38–39), which may give the impression that all grammars are somehow badly constructed. It would, however, be better to say that "The idea of grammar leaks" because the basic binary notion of acceptability simply does not fit the facts of any language. And that is actually what Sapir was talking about:

> The fact of grammar, a universal trait of language, is simply a generalized expression of the feeling that analogous concepts and relations are most conveniently symbolized in analogous forms. Were a language ever completely "grammatical," it would be a perfect engine of conceptual expression. Unfortunately, or luckily, no language is tyrannically consistent. All grammars leak.

Rather than talking about the universal linguistic features or structures that Tomasello claims have never been discovered, Sapir says merely that the fact of grammar is universal. And grammar for Sapir is a "generalized expression," not a fixed hierarchy of rules or structures but instead something that is never tyrannically consistent. His generalization is not susceptible to binary decisions of acceptability because, like many another generalization, it captures what usually happens without governing every case.

Complexity science and grammar

Complexity science, the complex system of language in use, tells us why this should be so, through a combination of nonlinear frequency distributions and scale-free networks. First of all, we have the 80/20 Rule. The different ways of putting words together have a nonlinear distribution, just like other linguistic features. Just to show once more the generality and relevance of the A-curve pattern, Michael Stubbs has reported the top-ranked collocates of *undergo* in a corpus of British English. He never claimed that this distribution was like Zipf, but it certainly has the shape of the curve when we graph his list (Figure 4.3). Of course the long tail of the curve is cut off here, since Stubbs only reported the most frequent collocates. Collocation is a different sense of "putting words together" than grammar per se, especially when we consider grammar to be composed of constructions as the usage-based linguists do, but it is difficult to know when the notion of collocations ends and constructions begin. Nick Ellis and Diane Larsen-Freeman have documented the nonlinear distribution of grammatical "constructions," as shown in Chapter 3. For Stubbs' corpus we cannot calculate the 80/20 proportion because we do not have all of

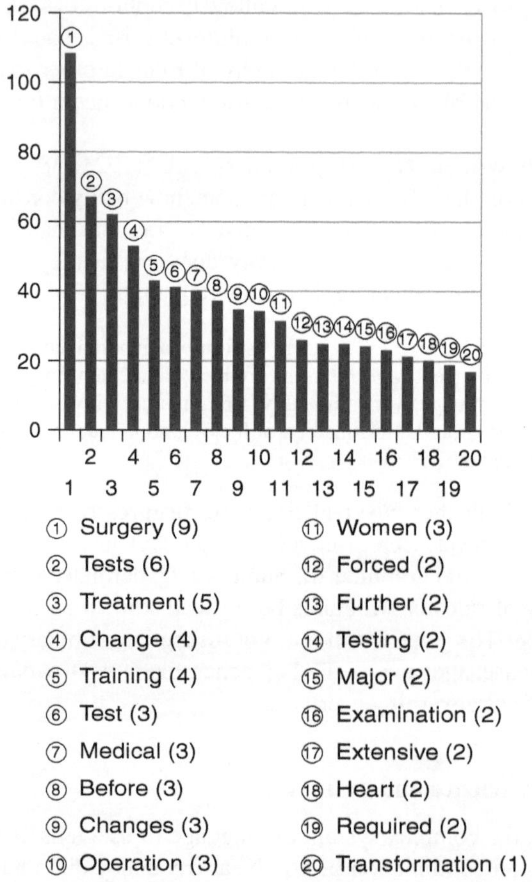

Figure 4.3 Stubbs, top twenty collocates of *undergo*, with frequency of collocate co-occurrences given in percentages in parentheses (adapted from Stubbs 2001: 89)

the data, but we can see that the top 20 percent of co-occurrences belong to the top three collocates, *surgery*, *tests*, *treatment*. The top twenty collocates occupy 63 percent of co-occurrences, which is many times more than we might expect if words were arranged randomly – Stubbs has calculated that a 1 percent co-occurrence rate for this corpus, the rate for the twentieth-ranked word here, is 125 times greater than would be expected by chance (2001: 73–74). Feedback in the complex system of speech has created a pattern, one that we see wherever we look in speech data, that is wildly improbable by chance.

What is important to note here is that the 80/20 Rule for such nonlinear distributions (whether the actual percentages are 80/20 or 90/10 or 70/30) tells us that we will always find one or a few constructions that account for the great majority of the instances for the feature under study, and that there will be a large number of variant constructions for the feature that account for a small minority of the instances. What the 80/20 Rule tells us about grammatical rules, then, is that we know in advance that there will be exceptions. Indeed, exceptions are not rare events, because we can predict that the class of exceptions will account for about 20 percent of the instances of any feature we study. Moreover, the exceptions will account for about 80 percent of the different constructions possible for any feature. Once we understand that the 80/20 Rule is not a curiosity but instead the hallmark of a complex system, we can understand what we take to be grammatical regularities in a different way. Grammatical rules, it turns out, are not laws but more like guidelines.

Grammatical rules are, however, more than mere suggestions because they have a nonlinear frequency curve behind them. Indeed, it is highly likely that the 80/20 distributional pattern gave rise to the idea of grammar in the first place, because speakers of any language perceive that for any question about how to put words together, there will be one or a few constructions that occur a great majority of the time, in the 80 percent group. The idea that there is a fixed objective language hierarchy, a linguistic system, originates as an observational artifact, something that we just *perceive* to be there because we usually do one of just a few things for any construction. Variants of features that are massively more common have a strong perceptual advantage because of the ever-present nonlinear distributional pattern, and they come to be perceived as "grammatical" or "acceptable." On the other hand, the less-frequent variants in the 20 percent group in long tail of the curve are at best perceived as exceptions, and at worst perceived as "unacceptable" or "ungrammatical." If speech is a complex system from which order emerges in the shape of nonlinear distributional curves for every linguistic variant, as all the evidence confirms, then grammar cannot begin as an underlying system of rules that generates all and only the acceptable structures. Grammaticality is something that we notice after the fact of usage, given our perception of the nonlinear frequency distribution of variants. Complex systems, then, support Paul Hopper (1987) when he called grammar "epiphenomenal," because our perception of grammar arises as a secondary effect of the complex system of speech, not as an intrinsic cause of how languages work.

It follows, then, that the 80/20 Rule that arises from the output of complex systems can account for Sapir's statement that "the fact of grammar" is "a universal trait of language." The Five Graces Group was wrong to make the tacit assumption of universal grammar, just as the generativists were wrong to do so. What is actually universal is the trait that speech is a complex system. Grammar is not a fact because it is an underlying system, but because of the *belief* of language users in the real existence of languages and dialects, based on their perceptions of the A-curves for every feature. That is, people share cognitive tools for processing their experience with speech, and the nonlinear distribution of variants allows for habituation of neural pathways for common features in the 80 percent group (whether or not we choose to call such habituation "mental representations" or instead consider it to be the output of neural networks or dynamical systems). We then come to believe that the variants with which we are most familiar constitute our language, and more locally constitute our dialects. The components of a grammar for some particular population of speakers are first and foremost the top-ranked variants of every linguistic feature, those in the 80 percent group, which linguists can make into a hierarchical system. Traditional grammarians also perceive that there is an 80 percent group of frequent usages, and they just substitute their own perceptions to make a grammar, instead of the scientific observations made by linguists of the 80 percent group in a community. Even though the nonlinear distribution of linguistic tokens is not something that our grammars have taken into account, it is the nonlinear distribution that allows us perceive it to be instantiated through rankings of feature variants for any and all populations of speakers. Grammar is thus, as Saussure has written, "a social phenomenon," something "social in its essence and independent of the individual," but for a reason that he had no way to anticipate. The idea of grammar requires, for its very existence, the nonlinear distributional pattern of linguistic variants that continually emerges, at all levels of scale, from huge numbers interactions between speakers who deploy feature variants in the complex system of speech.

Improved grammars

Let us now consider how our knowledge of speech as a complex system can help us to improve the grammars we create. First of all, the foundation of grammar in nonlinear distributions allows us to emphasize in contrast that prescriptive rules, from Saussure's dark side of grammar, have no such foundation. Instead of the justification of the ever-present A-curve of frequency, prescriptive grammars rely on someone's perception of social acceptability, whether the opinion comes from the Academie Française or, for English, from self-appointed usage mavens. Prescriptive judgments are thus inherently arbitrary in a way that grammatical descriptions based on the A-curve are not. The 80/20 Rule weighs against arbitrary judgments of acceptability and supports the idea in the *Comprehensive Grammar* that grammar is "immanent" in language – just not built-in in the same way that the authors thought it was. A knowledge of complex systems therefore frees us to consider grammatical prescriptions as a social affectation. Prescription is inherently political, a product of social debate and the exercise of social influence, in a way that the frequency profiles arising from the complex system are not. Judge Joiner's view that "standard English" belongs to the schools and to the professions can thus be seen as a political statement, as can Justice Scalia's belief in "plain language." The importance of Judge Joiner's Ann Arbor decision comes in its recognition that "black English" exists as a home and community language, not in its ratification of "standard English" as the property of the educated class. Justice Scalia's belief in "plain language" is revealed as an activist judicial tactic that justifies a judge's belief in his own perceptions of language, as opposed to historical or scientific modern views of how people actually use language. When we know the true grounds of prescriptive grammar, and that nothing about prescriptions is natural or inevitable, we can do a better job of managing expectations in our schools, where notions of prescriptive grammar are fostered and maintained. So, it is not true that we might as well be prescriptive because of the natural irregularity of grammar. We just need to use the 80/20 Rule to modify our understanding of how the "rules" of grammar arise, and to see more clearly how prescription differs from descriptive grammar.

Quirk et al., ever the most reasonable of grammarians, knew nothing of the A-curve and complex systems, but they did include a notion of frequency and its relation to acceptability. They were interested not just in binary decisions but in "regularities" (1985: 12):

> The study of words is the business of LEXICOLOGY, but the regularities in their formation are similar in kind to the regularities of grammar and are closely connected with them. [the authors provide an appendix on word formation]

They further specify that the regularities of grammar cannot be entirely separated from other systematic ways of observing the language, from phonology or lexicology or semantics or pragmatics, so that, they say (1985: 15),

> Our general principle will be to regard grammar as accounting for constructions where greatest generalization is possible, assigning to lexicology . . . constructions on which least generalization can be formulated. In applying this principle we will necessarily make arbitrary decisions along the gradient from greatest to least generalization.

Quirk et al. would not be very scientific if they were proposing that we should keep rules when they work and throw everything else into the category of lexical exceptions (unfortunately, this is a strategy not unknown in linguistics). However, the important point here is their recognition of "the gradient from greatest to least generalization." They are proposing a continuum in which grammar permits the greatest degree of generalization, but still not perfect generalization. Further to the same point, Quirk et al. create a relationship between acceptability and frequency. They claim that "Assessments by native speakers of relative acceptability largely correlate with their assessments of relative frequency" (1985: 33), and they have used the major corpora available at the time of their writing (Survey of English Usage, Brown, LOB) in order to establish the frequency of language phenomena. Quirk et al. thus differ from Chomsky's famous rejection of corpus evidence (see e.g. Andor 2004). They are also quite different in approach from Chomsky's principle of elegance in the creation of generalizations, which shows Chomsky's confidence in our ability to make binary judgments, a confidence not supported by the 80/20 Rule for language in use.

For the *Comprehensive Grammar*, since it already includes the idea of a continuum of acceptability based on frequency, all we need to do to make it align with the complex system of language in use is to say that any grammatical rule is subject to the 80/20 Rule. This change for the *Comprehensive Grammar* would really amount to a change in tone, not in substance. Instead of presenting rule systems that could be confused for prescriptions, the grammar would freely admit its leakage and build "exceptions" into the discussion of regularities – as the *Comprehensive Grammar* already does for some rules. The application of the 80/20 Rule would also allow us to evaluate how different constructions are included as part of a grammar. We can use frequency to determine whether any given construction is in the 80 percent group, and thus a part of what we might want to call the "core" of the grammar constituted by the top-ranked variants. Alternatively, we can determine that constructions are in the 20 percent group, the great majority of constructions that rarely occur, more like what Quirk et al. call "exceptions." To do so improves their structural grammar by specifying that their generalizations are actually guidelines: a grammatical rule, if well formulated in a structural grammar, can be expected to apply in the great majority of uses in practice, but we can expect a large array of infrequent cases where it does not. These are not all errors. The cases in the long tail of the A-curve are as much a part of the language as the highly frequent cases. The 80/20 Rule just gives a better metric for such evaluations, to replace the impression of some users of the grammar that rules are binary distinctions of what is right and what is wrong, and to minimize the necessity in Quirk et al. for the authors to make arbitrary decisions along the continuum.

For generative grammar, the best thing would be for Chomsky and others to continue to insist on the separation between language in use and the rational linguistic structure they prefer to talk about. As Tomasello has argued, the evidence of the world's languages has not borne out the prediction of representational innateness, and so generativists need not press that old case. Generativism, however, still offers a substantially elaborated model of logical relations in language, and we need not abandon that logic along with innateness. Indeed, we should not abandon a model that helps us to discover the linguistic value that

Saussure identified. The generative model begins with idealization in order to avoid the complex interactions of speech – the ideal speaker in a homogeneous community – and yet it must still have some empirical connection with speech. The 80/20 Rule suggests that there is no end to the problem of trying to write rules that can generate all of the acceptable sentences of the language. It cannot happen because of the long tail. Once we accept that infrequent constructions are normal parts of language in use, then we can understand that there are too many constructions to accommodate in any elegant rule system. It will, however, continue to be possible to write rule systems that account for the 80 percent group in the 80/20 Rule, and that is in fact what has mostly been happening already. Indeed, the 80/20 Rule can offer assistance to generative grammars because it also provides a metric for them, like Chomsky's earlier metric of elegance, to help decide what constructions to include in such a grammar and which ones to leave out. Application of the 80/20 Rule to a generative grammar would retain the core variants that are most important to linguistic value, and offers a principled way to set aside variants on the periphery. Bill Labov once famously said, in his essay "Logic of Nonstandard English," "For many generations, American school teachers have devoted themselves to correcting a small number of nonstandard English rules to their standard equivalents, under the impression that they were teaching logic" (1972a: 255). If we do not apply the 80/20 Rule to generative grammars, we risk turning Labov's statement around: generativists would be proposing a small number of rules of logic, under the impression that they were describing language. The 80/20 Rule offers generative linguists a better way to align language in use with the kind of binary logical decisions important to the generative model. And those binary logical decisions are just what is required to analyze linguistic value, as Saussure described it, the meaning that comes from hierarchical relations within a linguistic system.

The 80/20 Rule thus highlights an important difference between structural and generative grammars. Neither one tells the whole story of grammar. As its name says, the *Comprehensive Grammar* does the best job of collecting all of the grammatical constructions of English. The structural grammar is comprehensive at the cost of allowing exceptions, of including the constructions in the long tail. Generative grammars, on the other hand, are superior for the analysis of linguistic value because of their reliance on binary logic, but they retain this superiority only when they study just the top-ranked variants and do not attempt to be comprehensive. As consumers of grammars we want both kinds, and we want them to concentrate on their strong points. We do not want structural grammars that try to compete with generative grammars by asserting ironclad binary rules that instead come to look like prescriptions. We do not want generative grammars that try to compete with structural ones by accounting for every low-frequency construction, and so lose the ability to write elegant rules that reveal linguistic value in the core elements of grammar. We should prefer structural grammars that do the best job with paradigmatic relations, listing and organizing all of the available constructions. We should prefer generative grammars that do the best job with syntagmatic relations, revealing linguistic value in the relationships between constructions. To make Hugh Kenner's insight into a metaphor for grammars, we want all of the books in our structural library, but we find value in having the books we use most often on our generative desks.

Complex systems offer us another way to write better grammars when we apply the principle of scale-free networks to what we have learned from the 80/20 Rule. The property of scaling tells us that we have to pay close attention to exactly what population of speakers or texts our grammar will apply. Contemporary sociolinguistics addresses the evident problem that different groups of speakers, whether distinct according to geographical/social criteria or just at different levels of scale, have different language behavior and thus different grammars. Grammars can only be defined for the speech of one population at a time because, while the 80/20 Rule always applies, it will apply differently for every different population of speakers. There is literally an infinite number of possible grammars, because the number of possible groupings of speakers along the geographical/social continuum is infinite. Our problem with scaling is how to relate the grammars described for different population groups, either as part to whole or between different parts of the whole. If we use the same array of grammatical constructions to compare two or more populations, comparison of the results will show that some variant constructions will be higher and some lower on the A-curves in each population, and that some will be in the 80 percent group for one population while they are down in the 20 percent group for another population. We can measure these frequency differences, but they necessarily create categorically different grammars. There will be differences in which different variants happen to be top ranked for different populations, and also differences in whether particular low-frequency variant constructions happen to be present or absent for different populations. To use the terms already suggested, constructions in the "core" and "periphery" will therefore differ, so that in grammars for different populations, constructions will sometimes belong to the "core" and sometimes appear as exceptions, and the linguistic value from relationships in the hierarchy of constructions will necessarily be different. The same problem occurs for different kinds of texts. As Douglas Biber has demonstrated (1988; Biber, Conrad, and Reppen 1998), there is a continuum of texts in multiple dimensions that we can observe by measuring the frequency of an array of grammatical constructions. The notion that "every text has its own grammar" is true, for text types, in the same way that the language behavior of different groups of speakers yields different grammars. Complex systems explain how this could be so, and explains why we should avoid the proposal of the Five Graces Group that speech is directly connected to grammar and the consequent implication that a language has only a single grammar. Grammar is always relative to the population of speakers and texts that we examine.

Once we know about and expect the property of scaling, we can accommodate it in our grammars. For one thing, as I have argued in detail elsewhere (2009: ch. 7), we cannot rely on anecdotal evidence. Because of scaling, we cannot assume that others will share our own intuitions about grammatical constructions. To indulge in another metaphor, we cannot assume that speech is like the paint on a wall, where we can tell the color of the entire wall by looking at any small part of it. We need to use valid randomized statistical sampling when we measure the frequencies of constructions in different, particular populations. We can do so for very small populations, as sociolinguists now do for communities of practice: we can sample from just a small section of a wall, and tell what color the section is, say the part of the wall that has been behind a picture for many years. Alternatively, we can sample from an entire wall and tell what color the wall is, say, the wall in a room with an accent color. And, again, we can sample from the walls of an entire room and tell what color the room is, whether it is dark or light overall. The judgment we make about the room as a whole will not tell us about an accent wall or the darker patch where the picture used to hang, and vice versa, the small patch or the single wall will not tell us about the room as a whole. In some cases they might – there might not have been a picture above the patch we sample, and all the walls may be the same color. But in the complex system of speech not only can we not rely on such uniformity, we know even from our own experience, no sampling required, that English is continuously variable around us. If we take care to match the grammars we make to particular populations we can write better grammars, whether for English as a whole, for national or regional varieties, for social groups, or for particular kinds of texts.

The *Longman Grammar* by Biber, Johansson, Leech, Conrad, and Finegan (1999; also Biber, Conrad, and Leech 2002, from which the illustrations in Figure 4.4 are adapted), for example, makes an excellent start on the process of recognizing differential frequencies of different grammatical constructions in four different "registers," different text types, conversation, news, fiction, and academic prose. So, in Figure 4.4 different relativizers show up at different frequencies in the four registers; the chart for fiction looks most like an A-curve as the authors ordered the forms, but each chart here has been reordered to make a good A-curve. Conversation has a low frequency of relativizers overall, but the A-curve would appear more clearly there as well if the scale of the y-axis were changed to fit the data. In Figure 4.5 the same distributional pattern appears for negatives across the registers, at right for the frequency of function word classes within conversation and within academic prose, and for the syntactic

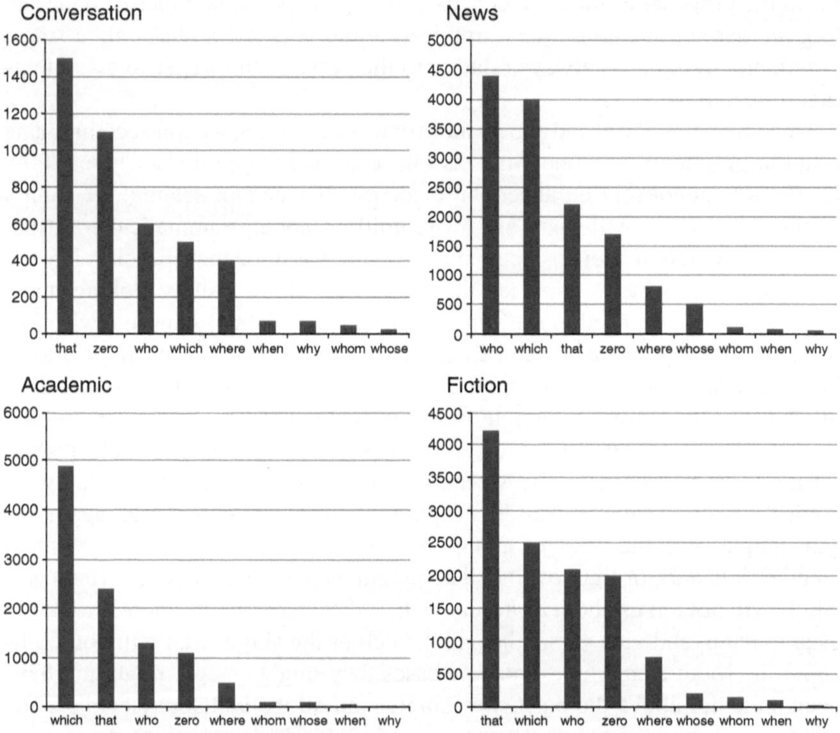

Figure 4.4 Differential frequencies of relativizers in Biber et al. the *Longman Grammar* 2002 (adapted from figures 8.14–8.17)

forms of adverbials in the overall corpus. In the latter case, the last category is "other forms of adverbials," so we know that the long tail would appear if these forms were broken out separately. These illustrations make it clear that, when we associate the frequency of constructions with text types or populations of speakers, we will not be able to generalize from individuals to small groups to larger groups of speakers or texts, or across parallel populations, so we will always need to be satisfied with what our sampling tells us for the particular population under study.

What, then, of the idea of Standard English, from the perspective of complex systems? Standard English is an institutional construct, and not, to use Quirk et al.'s term, something "immanent" in the language. The idea of Standard English is of a different order from the grammars that we can write based on the frequency of variant constructions from particular populations. Because Standard English is institutionalized, it promotes variants as if they were top ranked, even though they may not in fact be top ranked across a national

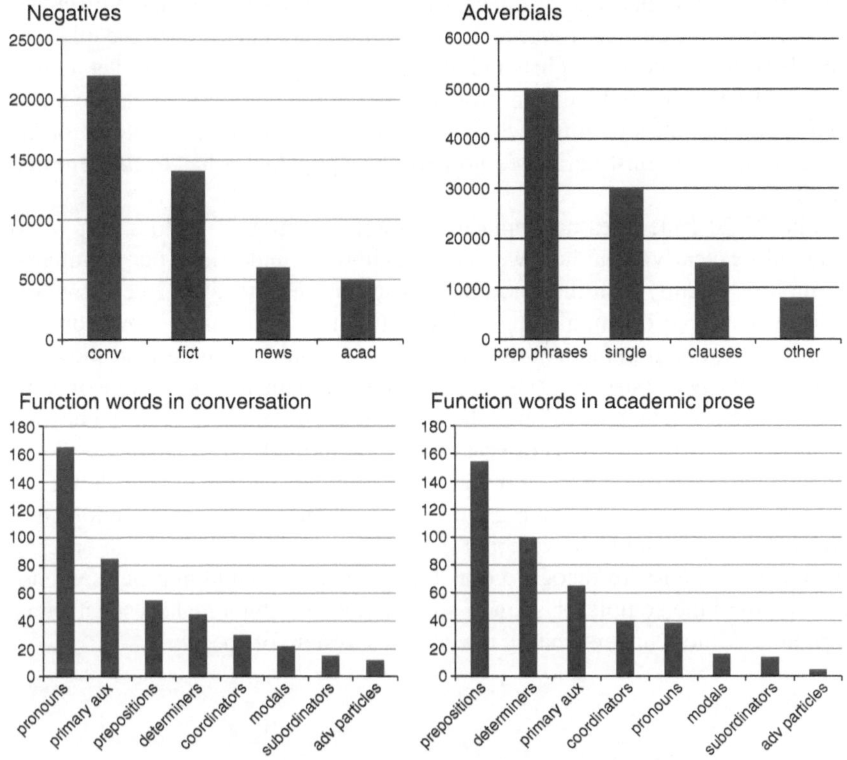

Figure 4.5 Differential frequencies of various constructions in Biber et al. the *Longman Grammar* 2002 (adapted from figs. 2.2–2.3, 8.6, 11.2)

population or for the group of English speakers as a whole. In this way, Standard English is on the edge between prescriptive grammar and descriptive grammar. The variants promoted institutionally in Standard English are often presumed to be those of some preferred population of speakers, but in fact experimental speech evidence shows that highly educated speakers (the group typically assumed to embody Standard English) vary substantially in their language behavior, whether between countries (British vs. American Standard), or between regions, or between localities, or even between different kinds of occupations. In contrast to Judge Joiner's formulation, it is simply not true that "the school, the commercial world, the arts, science, and professions" all share the same grammar. Artists do not talk like lawyers. Scientists do not talk like businesspeople. What the schools try to teach is not identical with the usage of any post-secondary educated community. Unlike the social arbitrariness of prescriptive grammar, there are good reasons to impose institutional standardization, such as spelling. Any agreement that limits the variation in grammar that will necessarily arise across time or large distances will improve the chances for successful communication. However, the relationship between Standard English and the usage of any individual speaker, or any group of speakers, will remain probabilistic. Any claims for linguistic value extracted from the structure of Standard English would be relevant just for the institutionalized construct, since no particular population has been identified for it.

The 80/20 Rule does not replace traditional grammars. It does help us to think more clearly about how we make and how we understand our grammars. Complex systems, with their nonlinear distributions and scale-free networks, offer a means to distinguish prescriptive traditional grammar from contemporary scientific grammars of different kinds. Complex systems also shows us how to make better structural and generative grammars. If a knowledge of complex systems can help us to relax the sometimes strident competition between promoters of different kinds of grammars, that too will be welcome. Once we understand the relation of a grammar to the complex system of speech, we are free to do a better job of making the kind of grammatical generalizations that we will always want to make about a language, and need to make in order to exploit what Saussure recognized as linguistic value. Grammar per se is not the output of the complex system of speech, but we cannot make a satisfactory grammar without understanding the complex system of speech.

5 Complex systems and the history of the English language

Cambridge is one of those old places that seem never to change, and yet there is a new Corpus Clock right in the middle of things, on an outside corner of the staid Corpus Christi College (founded in 1352) and right across from the sixteenth-century King's College Chapel, famous for its Christmas Eve choral broadcasts (Figure 5.1).[1]

Stephen Hawking, who wrote *A Brief History of Time* (1988), had unveiled the clock in 2008 and thus added a bit to time's narrative. The rather disturbing "chronophage," or 'time eater,' atop the clock eats the hours and the seconds as they pass. Its insect-like form was perhaps suggested by the name for the workings of the clock, a "grasshopper escapement" originally invented in the eighteenth century and here enhanced by an electric motor that winds a spring to keep the clock in motion. The other disconcerting thing about the clock, after its striking appearance and its equally amazing appearance at all in old Cambridge, is that it is only accurate every five minutes: time slows down for a while, even stops, and then speeds to catch up. Of course this trick is on purpose, something that, like the chronophage, is meant to unsettle us from our comfortable assumptions about time and its regular passage. Thus Stephen Hawking was the right person to unveil it, owing to his work in contemporary physics that challenges our assumptions about time and the universe around us, and thus too a fitting emblem of the new ideas from complexity science as they affect the field of historical linguistics. This chapter will take up the topic of language evolution from the point of view of the Corpus Clock, that the evolutionary process can speed up and slow down in time, rather than proceed in regular steps. It will then suggest some practical applications of complex systems to historical linguistics, things that we can do now to improve how we think about our work. The examples will come from English, but at the same time it should be clear that the principles described are not limited to English and can be applied to any language.

[1] This chapter is based on Kretzschmar (2015).

Figure 5.1 The Corpus Clock (from http://deetoursbydeepali.wordpress.com/2012/05/24/the-time-eater-of-cambridge, used under Creative Commons Attribution-ShareAlike 3.0 Unported license)

Clockwork

Let us begin with a linguistic analogue of the Corpus Clock: a brief historical analysis of the word *clockwork* in Mark Davies' American English corpora (Figure 5.2; http://corpus.byu.edu). In his contemporary COCA corpus (450 million words from 1990–2012, accessed April 2014), the word *clockwork* occurs 493 times, and it is accompanied by these collocates within a span of four words left and right. The list clearly follows the nonlinear curve we expect of a complex system, though the long tail of single occurrences is cut off in Figure 5.2. We can set aside for the moment the collocation "clockwork orange" as the creative title of a popular book and movie. The largest use of *clockwork*

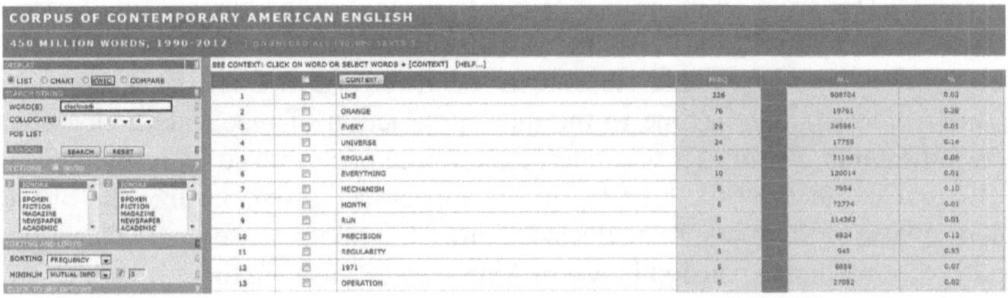

Figure 5.2 Collocates of *clockwork* in COCA

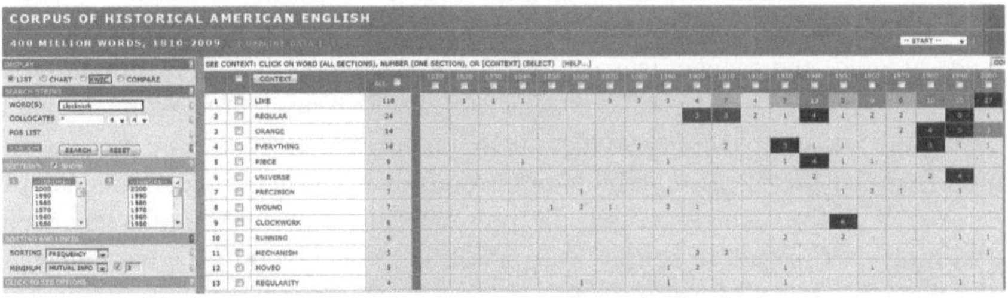

Figure 5.3 Collocates of *clockwork* in COHA

comes with *like*, and with words such as *every* with a time unit (as in "every month like clockwork"), or *everything* (as in "everything runs like clockwork"). Similar to such usages, we have *regular* (as in "regular as clockwork"). These are all similes, marked metaphorical uses that understand *clockwork* to mean the epitome of control. Perhaps the title "clockwork orange" plays creatively with this meaning as well. The same might be said for the unmarked metaphorical use, "clockwork universe," which understands the universe to operate by Newton's laws as regularly as a clock runs. As against these metaphorical uses, we have *mechanism*, which refers to the literal wheels and gears of a clock or of a machine like a clock. The point here is to show that our contemporary understanding of *clockwork* is not so much the literal sense, although we do have it still, but instead the metaphorical sense of regularity, whether in daily events or in the universe itself. Time passes at a constant rate, as measured more or less precisely by clockwork. If we compare these results to Davies' historical COHA corpus (Figure 5.3; 400 million words of American English, collected by decade over two centuries), we see that *clockwork* occurred 353 times. The greater number of tokens in the contemporary corpus of about the same size from the historical corpus must mean that its rate of occurrence has increased significantly in recent decades, ironically during the time when digital clocks have been replacing clocks with physical mechanisms. Mechanical clocks were invented in the Middle Ages, and yet the first citation of *clockwork* in the *OED* is from 1628, in its metaphorical sense in the phrase "this curious clocke-worke of religion." The literal sense is only documented in *OED* from 1662. The rise in frequency of *clockwork* in American English as shown in COHA corresponds in time to the mass production of mechanical clocks in the nineteenth century (it was one of the initial such industries in the USA). The unusual facts of the time lag in the appearance of *clockwork*, and of the appearance of the metaphorical sense before the literal sense in *OED*, and its relatively sudden expansion in use in American English are topics to which we will return shortly.

In the COHA list we again see the nonlinear curve we expect from complex systems in the distribution of collocates, with many of the same words. However, we do see the operation of historical change. In the nineteenth century, *piece*, *wound*, and *mechanism* are much more common than the metaphorical uses with *regularity* and *precision*. The typical meaning of *clockwork* changed during the twentieth century so that the metaphorical sense became more common than the literal sense, which can account for the increased frequency of *clockwork* at the same time that the presence of its literal referent has been declining. Both of these senses were possible in each century; what changed was their frequency of use.

If we look a little closer at *clockwork universe*, however, between its emergence in COHA in 1940 and its greater prevalence in the recent period covered by COCA, we see that almost all of the examples are used in a negative sense: they refer to the demise of seventeenth-century Newtonian ideas of a regular, controlled universe in favor of the relativism of Einstein and Hawking. So, for example, people are writing things in reference to quantum mechanics like "That new theory replaced the perfectly predictable, **clockwork universe** of Isaac Newton with a mysterious world" (COHA 1980s token). So, in another historical development, it appears that in the most recent time period, we have come to reject the *clockwork* metaphor when it refers to our universe. In a particular domain, in usage at one small scale, we find that the meaning of the word does not follow the frequency pattern that it does more generally at the time. Again as we would expect of a complex system, we need to take scale into account in our observations, to understand that linguistic behavior may be different in different subsets of our data, in different situations of use. Besides the matter of scale, contingencies may arise to affect the pattern, such as the sudden innovation "clockwork orange" also at the end of the twentieth century.

Furthermore, if we do see different linguistic behavior in different places or in different text types, nothing has gone wrong; such differences are not evidence of some error by the writers or of some error in our analysis. *Clockwork* not only has different frequencies of its meanings over time, but those meanings occur at different frequencies in different domains at the same moment in time. We now see that, because of the operation of speech as a complex system, and in line with our modern view of a "clockwork universe," this semantic aspect of the history of the language does not operate like clockwork. It took too long, we might think, for *clockwork* to appear in American English, and the word has had a complicated and changing semantic history. Ironically, the history of *clockwork* shows us how time appears to speed up and slow down, on occasion even to stop (in the delayed appearance of the word), when we inspect the continuous history of the word's semantic development.

Grammaticalization

Now let us consider time over the long term not for semantics but as regards the topic of grammaticalization in language change. The origin of the idea is Paul Hopper's influential 1987 paper, "Emergent Grammar," as described in Chapter 3. For historians of the language, the key question is how to accommodate Hopper's sense of continual movement, and still be able to describe the grammar of the language at any moment in time. Grammaticalization came up in Chapter 3 as just one component of usage-based linguistics, and it is appropriate now to describe its mechanics in more detail. The Five Graces Group cited in Chapter 3 (Ellis and Larsen-Freeman 2009a: 1–26; originally published as Beckner et al. 2009) begins with a basic tenet of grammaticalization (6):

> Historical changes in language point toward a model in which patterns of co-occurrence must be taken into account. In sum, "items that are used together fuse together" (Bybee 2002). For example, the English contracted forms (*I'm, they'll*) originate from the fusion of co-occurring forms (Krug 1998). Auxiliaries become bound to their more frequent collocate, namely the preceding pronoun, even though such developments run counter to a traditional, syntactic constituent analysis.

They continue to specify the goal of their position (2009a: 7):

> In the usage-based framework, we are interested in emergent generalizations across languages, specific patterns of use as contributors to change and as indicators of linguistic representations, and the cognitive underpinnings of language processing and change.

For the Five Graces, emergence is the same thing as grammaticalization. *I'm* and *they'll* both emerge and grammaticalize at the same time. The example of the emergence of English contractions gives a simple shape to more complex "generalizations across languages." Grammaticalization, then, becomes for the Five Graces a universal human process, one which describes the operation of the complex adaptive system of speech. Grammaticalization is the output of the complex system, what happens as a result of the process of interaction. It is taken to operate at the level of single languages, but because it is universal it is possible to look for generalizations across languages. Since the Five Graces propose that grammar is "a network built up from the categorized instances of language use," then grammaticalization is the mechanism by which they propose that such a network arises and changes. As we have seen in Chapters 3 and 4, this kind of mechanical change is not what complex systems do. Like Hopper and Traugott's 1993 book, the Five Graces retreat towards the traditional mainstream of linguistics and elevate grammar to an essential part of language while, as we have seen in Chapter 4, the grammars created by linguists can only be perceptual artifacts based on the nonlinear distributional pattern, the 80/20 Rule, that is actually the output of the complex system.

Let us consider further the "network built up from the categorized instances of language use," as the Five Graces describe grammar. This network can be seen as a kind of object that converts Hopper's process into a state, and identifies the state of the complex system as its grammar. In the language of complex systems, a "network" is a set of nodes, the elements in the complex system, without regard to the condition of each node. A "state" is the condition of all of the elements in the system at one moment in time, or of a single element at one moment. For a single traffic light (to borrow an example from Larsen-Freeman and Cameron 2008: 32), there are three possible conditions, red, yellow, or green, that the light could have at any moment in time. In the language of complex systems, the "state space," the set of possible states in the system, of the single traffic light consists of those three states. If we had two traffic lights, the state space would have 9 states (3^2 states: RR, RY, RG, YY, YR, YG, GG, GR, RY). If we had three traffic lights, the state space would have 27 states (3^3 states), and for four traffic lights it would have 81 states (3^4 states). Thus the problem for urban planners who are trying to organize traffic by trying to synchronize a great many traffic lights: every additional light in the plan causes a geometric increase in the state space, which in practical terms makes the problem unresolvable. And thus the problem for speech, which has many more elements than a traffic light and many more speakers in interaction than any city has traffic lights. If, as we have seen in data presented in earlier chapters, there are 25, or 50, or 100 possible states for a vowel or for a lexical cue, and we have a population of speakers of, say, 200 people (the median number of Facebook friends for an adult user in 2014), then the number of possible states in a network for the vowel or lexical cue is 25^{200}, or 50^{200}, or 100^{200}, any of which is a number much larger than we can imagine, many orders of magnitude larger than the number of stars in the universe (approximately 10^{24} stars, according to the current consensus estimate from various online sources such as the European Space Agency; NASA suggests a smaller number online, only 10^{21}). For the grammar of a language the problem is that much worse, because we should consider all of the speech sounds and words and constructions, not just a single cue. Moreover, for speech it is not clear that we can characterize the network of nodes because the categories for analysis such as words or constructions, unlike traffic lights, are not naturally given as discrete; we can identify whatever units we choose from the stream of speech to count in our analysis, but the units are essentially arbitrary and will overlap many other potential units for analysis. Given these problems of numbers and categorization, the grammar-as-network proposal of the Five Graces is simply untenable. To say that there is a "network of categorized instances of language use" amounts to the same thing as saying that "matter is made up of a network of atoms" or "the universe is made up of a network of stars," which may be correct in some sense but which can hardly qualify as a useful statement about networks. Referring to grammar as a (cognitive) network of nodes is thus a much less satisfactory model than the neural network or dynamical systems theories now available in cognitive science, as mentioned in Chapter 6.

For any single linguistic feature we choose to analyze, which is again not so simple as a traffic light because it has a great many variant choices instead of just three, the Graces refer to the process of "selection" as the emergence of a preference for one state of the feature, one variant, over other possible variants (Ellis and Larsen-Freeman 2009a: 9):

> Selection is a process by which individuals – organisms, or humans as speakers or cultural beings – interacting with their environment cause replication of the [units] to be differential; that is, some [units] are replicated more than others.

They associate change with long-term alteration in social practices (Ellis and Larsen-Freeman 2009a: 9), "which in the extreme case leads to the fixation of [new forms] and extinction of [old forms]," "to the rise and fall of words and constructions associated with those lifestyles (e.g., the rise of *cell* [phone] and the fall of *harquebus*)." The preference for selection, or "fixation," of a state takes the dynamic movement of the complex system and freezes it, so that one variant of a feature becomes "grammatical" in the sense of having been selected. The Five Graces' whole idea of selection thus runs against Hopper's idea of the process of emergence, and converts it into a single state of the network of nodes. And by the numbers, again, such "selection," if it were ever to occur, could not create any sort of usable network. If we were to consider that "selection" operated to identify the top 20 percent of variants for any feature, then we would still be faced with numbers like 5^{200}, 10^{200}, or 20^{200} for the state space of the network, numbers still greater than the number of stars in the universe. Of course there is nothing at all wrong with making grammars. This kind of formal analysis has been highly productive. But there is an essential conflict between making a grammar as a static hierarchy for a language at one moment in time, and the process of change in language that is best described as frequency change within a complex system.

The best answer from complexity science, again, is to observe the nonlinear distribution and scaling properties of complex systems, and to put the 80/20 Rule to good practical use, and these notions lead to suggestions for some improvements in our analytical practices for historical studies. The idea of selection should be reserved for rule-based generative grammars that need to "select" the most frequent forms of constructions in order to preserve the elegance of their logical systems. Generative grammars can afford to, and therefore should, ignore the 80 percent of possible constructions that occur rarely in order to concentrate on the 20 percent of the construction types that account for 80 percent of the tokens. To do otherwise results in "rule creep," the inelegant addition of more and more rules to account for infrequent cases, and thus also the loss of what best distinguishes generativism from structuralism. On the other hand, in structural grammars that collect paradigmatic lists of possible constructions, there is no good linguistic reason to privilege the most common variants as having been "selected" and therefore have status as being "grammatical" and to relegate less-common variants to "noise" in the system. It will be enough to note whether any particular variant is in the 80 percent core or the 20 percent periphery of constructions. All of the variants on the A-curve, both the top-ranked variants and those in the long tail, are actually just as relevant for inclusion in the structural system. The notion of language-internal selection, then, cannot operate within the A-curve because nothing is really chosen or preferred. For the historian of the language, this means that "change" can no longer be defined as a different variant having been "selected" at a later moment in time.

A-curve frequency profiles

If we observe the distributional pattern of linguistic features at any moment in time, we can get more historical information than we might have thought. Paul Hopper raised just this issue in 1987 in "Emergent Grammar" using the example of the development of the English indefinite article *a*/*an*, as presented in Chapter 3. English has preserved the historical functions of *a*/*an*, as well as developing new ones. Whether in proverbial language or in special domains like law or religion, we do still employ old constructions, which gives them a low frequency on the modern A-curve. Thus the extinction of forms is not an extreme case, as the Five Graces suggest, but instead a most unlikely one: even old forms like *harquebus* continue to exist at low frequencies, whether in historical accounts or, sometimes, in proverbs such as "Hoist with his own petard" ('to be blown up with his own bomb'). Particular domains such as religious language can carry over centuries-old morphology, as does the King James Bible. The long tail of the A-curve is also a perfect place for innovations in language, which can arise at low frequencies and which may or may not ever increase in frequency over time.

The indefinite article is therefore not an exception but a good example of a general capacity of speech to retain old feature variants at low frequencies.

One of the points made by C. S. Lewis in his *Studies in Words* (1960) is that the meanings of key words in our culture have "ramifications." That is, words do retain their older meanings even while they gather new ones. This fact may be obscured for incautious readers by what Lewis called the "dangerous sense," the meaning that is so frequent in modern usage that we automatically think of it. Lewis could have had no idea of A-curves, but the dangerous sense is of course the top ranked meaning on the modern nonlinear curve of meanings for a word. Our corpus analysis of *clockwork* showed us this process in operation in contemporary time: we can still use the word to refer to gears and wheels, and the dangerous sense of the word is regularity and precision, but we need to be aware that now we often speak of *clockwork* in a negative sense, as with *universe* and *orange*, that may give the word a dated or unsavory or ironic semantic prosody that refers to something a little too perfect in operation. Thus, for grammatical constructions, the lexicon, and semantics, all of the evidence suggests that historical forms tend to be retained as low-frequency variants in the tail of the nonlinear distribution of contemporary usage, and that such usage continues actively to change.

For historians of the language, this means that "change" in the complex system of speech will consist of an alteration in frequency of any particular feature, instead of the selection of one form over another. One way to track feature frequency, the S-curve, has already been described for the progress of linguistic change (notably in Kroch 1989; Labov 1994: 65–67), and the A-curve distribution is in no way at odds with the S-curve. The two curves are actually different expressions of the same basic distributional facts. The S-curve just describes the successive frequencies of a single variant at different moments in time. In Figure 5.4 we see two different A-curves that correspond to different moments in time for the same variant, and locates the position of the variant on each curve. We can recall, for instance, the changing frequencies of the meanings of *clockwork* over the past two centuries. If we then draw A-curve charts that correspond to different times on a given S-curve chart (Figure 5.5), we can see that the characteristically sudden positive movement of variants on the S-curve is mirrored by the shape of the A-curve. When a particular variant "climbs" the A-curve by moving up in frequency rank, the distributional patterns of both the A-curve and the S-curve predict that there is a larger relative change in the middle of the curves. As a variant moves from one rank to the next in the middle of the curve, each step in rank describes an increasingly large number of occurrences. This means that the small number of occurrences when a variant begins to become more frequent will be expressed as a slowly growing proportion, and the larger number of occurrences in the middle of the curve will be expressed as a rapidly growing proportion. The A-curve and the S-curve are thus complementary descriptions of the distributional facts of variant linguistic forms at different moments in time, new and improved

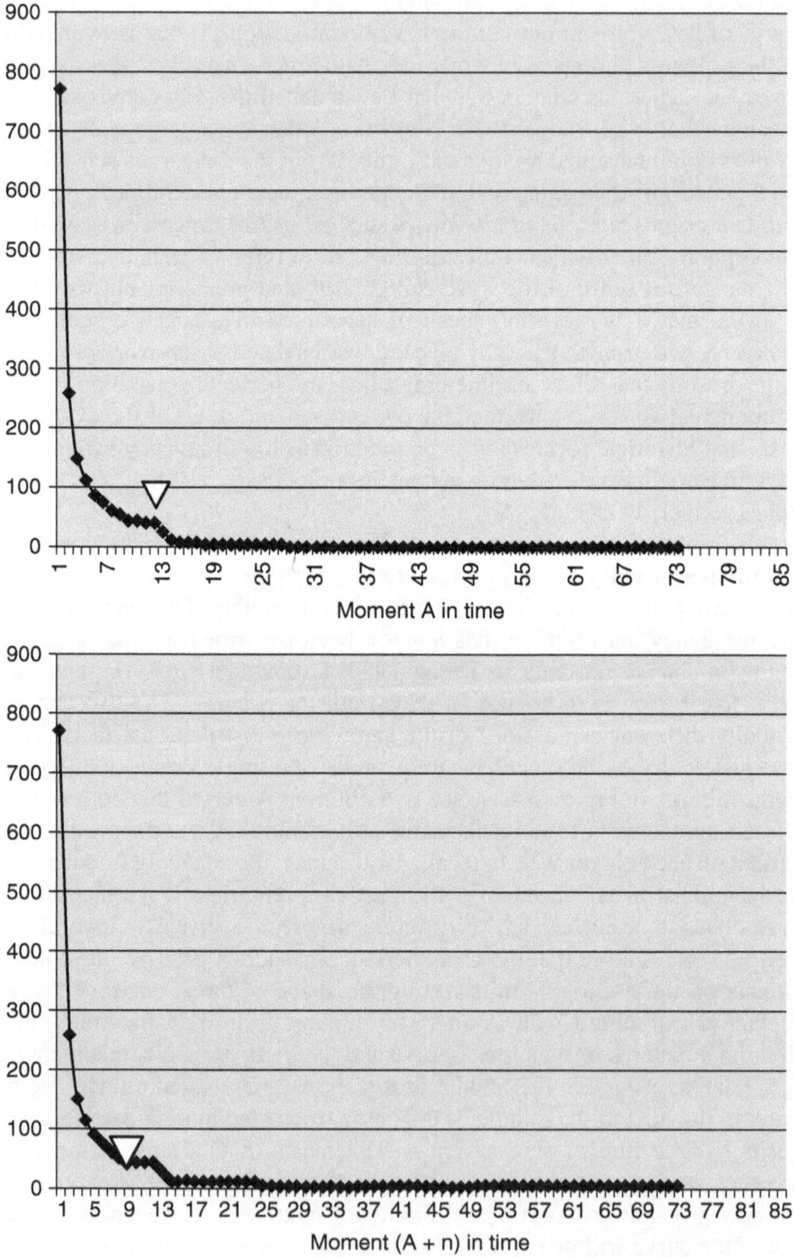

Figure 5.4 A-curves at two hypothetical moments in time (from Kretzschmar 2009: 266)

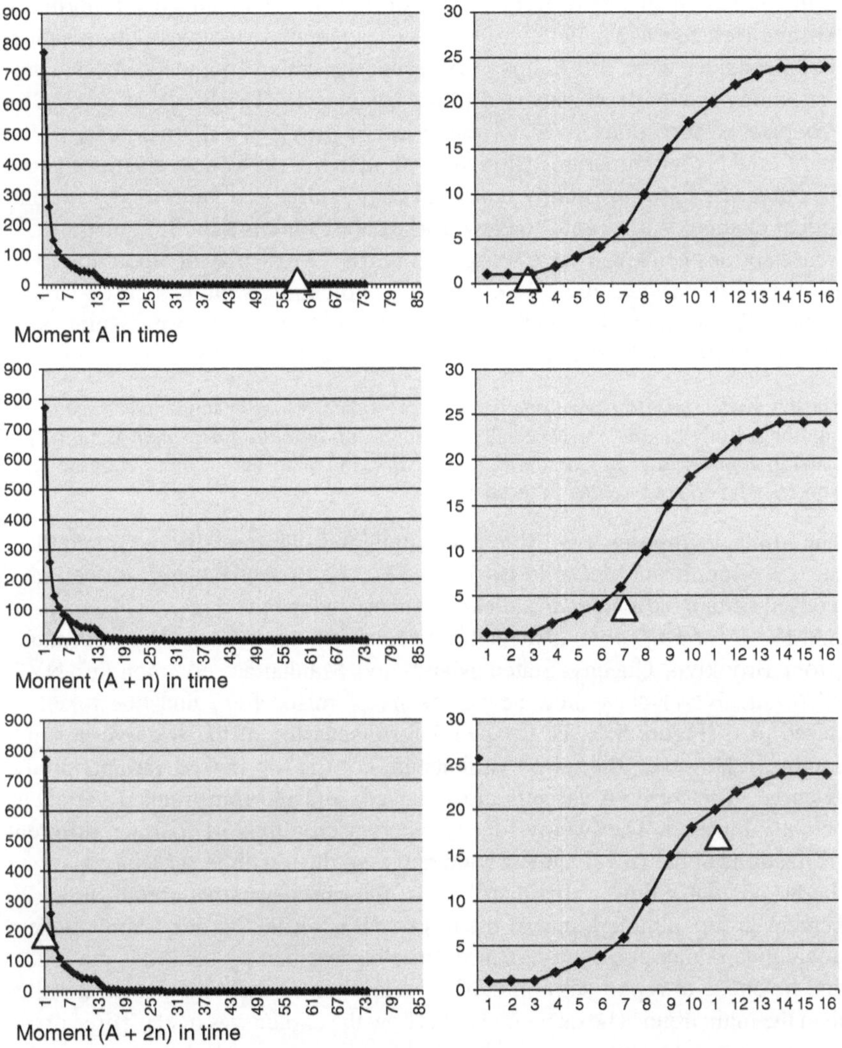

Figure 5.5 A-curves at different hypothetical moments in time, with associated S-curves (from Kretzschmar 2009: 268)

ways to document historical change. And this focus on frequency distributions, rather than qualitative change, also allows explicitly for what Laura Wright (2000: preface) has waggishly called "W curves" that describe increases and decreases in the frequency of the same form over time.

In contrast to change of frequency, historians of the language still normally talk about change as a qualitative event. As an extended example of this practice, let us consider what Bill Labov (1994) reported about Linguistic Atlas material among the evidence reported in his Chapter 4, "The Study of Change in Progress: Observations in Real Time."[2] Labov provided a sharp defense of the traditional Neogrammarian historical method in his 1994 book, and throughout his career he has consistently defended chain shifts as a valid model for historical change. With regard to New York, Labov claims that "Linguistic Atlas transcriptions of New York City speech in the 1940s showed a notation [æ┴] for tensed /æ/ in *half, pass, last, bad*, etc., indicating a slightly raised variety of [æ] (Kurath and McDavid 1961)" (1994: 74). As it happens, Kurath and McDavid (1961) talked about [æ] in New York City as follows (15):

> The low vowels /æ, ɑ/ are positionally and/or prosodically long or short, and occasionally ingliding. Long [æ·, æ·ᵊ] is especially common in *bag, half, glass, aunt*. Long [ɑ·] is usual in *John*, regular in *glass, bath, aunt, can't*, in the speech of those cultured New Yorkers who do not have /æ/ in these words . . .

The idiolect synopses, two from Manhattan and one from Brooklyn, confirm the text (Kurath and McDavid 1961: 55–57). The Brooklyn synopsis does have a raised variant of [æ] in *bag* and *glass*, and my independent review of data from LAMSAS for *half, past, last* for the twenty-four New York City informants (from Brooklyn, Queens, Staten Island, and Manhattan, communities NY5, NY6, and NY7) does show seven tokens of raised [æ^] and one token of raised [ɑ^] (Figure 5.6; see Chapter 7 for discussion of the A-curve in small samples). However, the raised variants are not the top-ranked variants on the A-curve. The top two variants are unraised, and the third-ranked variant is actually lowered. The twenty-four speakers used a total of fourteen different realizations of the vowel, quite a wide range, so the top three variants constitute the 80 percent group in this distribution; the raised variants are all in the 20 percent group, although one of them, [æ^·ᵊ], is on the border. Labov made a qualitative choice for the moment in time represented by the Atlas materials, and he chose to report some minority variants from the Atlas material rather than the main trend. The question here is how the choice was made. Three of the four speakers (of twenty-four) who used [æ^] were among the younger Atlas speakers (39, 43, 45), while the other was 60 years old. Still, two other younger speakers in New York (38, 48) did not use the raised variant, so it is not the case that all of the younger speakers in New York had raised variants. Speakers with the raised variant were also mixed for other social factors tracked by the Atlas (sex, education, occupation). One plausible explanation is that Labov only needed to see some evidence of the raised variant at the time of the Atlas

[2] This section is based on Kretzschmar (1996a).

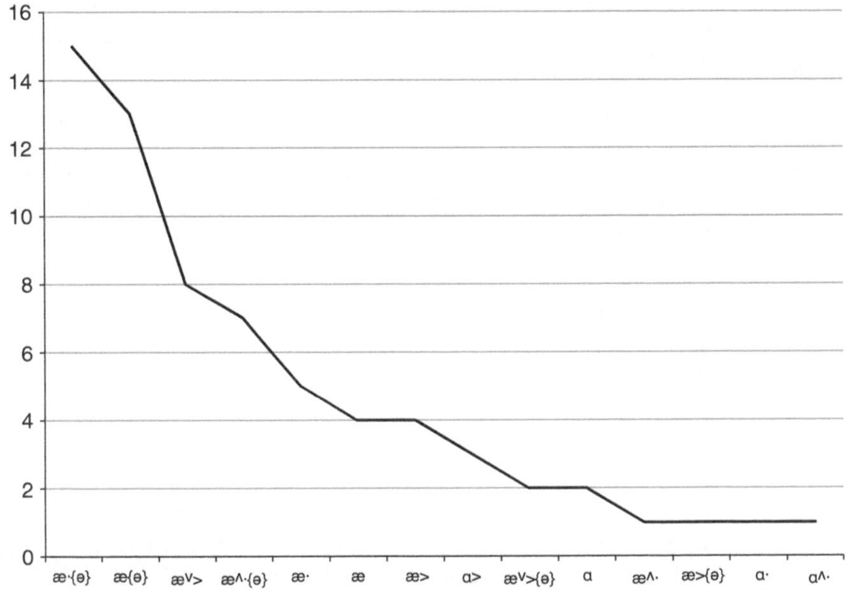

Figure 5.6 Realizations of [æ] in *half*, *past*, *last* from twenty-four New York City speakers in LAMSAS

evidence as part of the pathway of the change; he did see some evidence, and he reported it. Since the traditional Neogrammarian paradigm preferred to see a qualitative statement of the character of the vowel at that time, Labov provided one. Now that we know about complex systems and nonlinear distributions of variants, it no longer makes sense to make a qualitative choice.

Labov's report of Hubbell 1962 [1950] is similar, when he says only that Hubbell "observed the homonymy of *bad* and *bared*" in recordings made of Atlas informants (1994: 74). Hubbell does mention the homonymy of *bad* and *bared* (1994: 76), but only in the midst of a list of "variations in New York pronunciation" and not as his general characterization of the vowel (1962 [1950]: 59–60):

The vowel in *cap*, *back*, *match* is intermediate low-front. It normally has a somewhat lower and more retracted position that the corresponding Southern British vowel and, when it varies, does so toward a still lower and more retracted placement, rather than toward [e] as in Southern British... The phonetic distribution of /æ/ in the usage of many metropolitan speakers, particularly but not exclusively on the lower levels, differs markedly from that found in certain other regional types of English. Only under certain phonetic conditions does /æ/ occur in everyone's speech; in other positions it is replaced in the pronunciation of many, regularly or sporadically, by the phonemically distinct /εə˘/ (or /æə˘/).

Labov also reports that Babbitt 1896 "indicated /ɛ:ᵊ/ for this class" (1994: 74). Actually, Babbitt was particularly addressing the speech of "the lower classes in New York City," and did not use modern phonetic notation (1896: 461):

> *á* is very high, pretty close to *e* of the normal scale, and never mixed – being thereby clearly distinguished from the New York *e* (>ə). Among the older New Yorkers this very high vowel is used in all the set of words pronounced in New England with the broad vowel (*ask*, *half*, *pass*, etc.), and is really higher in these words than in *man*, *cab*, etc. But this distinction is now lost and the general vowel has quite overtaken the special one (*hend* hand, *keb* cab, *dens* dance, *hef pest* (half past). [*sic*]

Babbitt does say in his section on "Vowels with r" that the *-r* for his lower-class New York speakers is either "silent" or represented by a schwa glide (1896: 462–63), but says nothing about a glide in other environments. Labov filters the observations of Babbitt and the detailed evidence of the Linguistic Atlas and Hubbell to arrive at the following formulation:

1895 /ɛ:ᵊ/ | 1940 /e:ᵊ/ | 1960 /i:ᵊ/

To create a linear shift like this certainly requires a specialized view of the evidence. Length and a schwa glide have been added to Babbitt's (most likely) [ɛ], presumably for consistent appearance of the notation of the shift; Atlas [æ·ᵊ] and Hubbell's /æ/ or /ɛə̆/ or /æə̆/ have been represented as something higher, which in even the most favorable view would be an occasional variant as reported and not the preponderant form, presumably to serve as an intermediate form between Babbitt and Labov's 1960 observations. When we add Labov's treatment of Babbitt to what we have seen from the Atlas evidence, it does seem that Labov only needed to see some evidence of the raised variant at each time period as part of the pathway of the change; he thought that he did see some evidence, and he reported it. The traditional expectation that there should be a qualitative state for a linguistic feature at a given time has allowed Labov to be highly selective in what state he reports for each time. Labov's drive towards abstraction, to make generalizations in keeping with the Neogrammarian program, must necessarily color his observation of real-time evidence and its difficulties of interpretation.

In an attempt to address the difficult evidence, Labov asserts that "Dialectologists in the field are often quite conservative in their notation, and they tend to limit their entries to variants of the forms they have encountered before, even when the sound changes have advanced across several levels of the sound system" (1994: 74). This explanation suggests that the range of transcriptions should be much more limited than they are; Atlas fieldworkers should not have reported fourteen different realizations of the [æ] vowel if they were limiting their entries to what they expected to find. We do not have Babbitt's evidence, but Labov has reduced even what Babbitt reported in his summary to a single qualitative statement. The central fact here is that Labov's formulation of this New York City change radically reinterprets the evidence and finds a chain shift when the direction and progress of the change are far from clear. The evidence might be taken to support not an advancing chain shift, but no more than maintenance in 1940 of Babbitt's change among the lower classes: Hubbell found some raising "particularly but not exclusively on the lower levels" while he expressly denied it for majority use, although it is unclear whether the distinction from majority use that Hubbell wished to draw there was Babbitt's raising or the presence of the schwa glide that Babbitt only reported for words in *-r*. Atlas field work found no more than [æ^], never [ɛ], but some tokens of the slightly raised variant occurred at all social levels, so no subsection of the sample can be said to be leading the chain shift. At least in this case, and by extension for a great many other uses of qualitative statements about linguistic features following the methods of traditional historical linguistics, there can be no defense for such qualitative treatment of the evidence. The need to make a qualitative statement about a feature forces the analyst, even someone like Labov at the top of the field, to choose what evidence to present and what to leave out. The traditional method offers no guidance about how to make the choice, and so it is most likely that the cart will drag the horse, that the hypothesis of the analyst will control how the evidence is handled. Traditional historical linguistics allows too much subjectivity in analysis of difficult data.

The A-curve, on the other hand, offers historians of the language a way to judge their data when it is too sparse to construct its own frequency profile. Historical studies often cannot acquire enough data for a good survey, and the data that is available does not constitute anything like a valid sample of the speech or texts in use at the time of the study. Still, the 80/20 Rule offers a metric to estimate the status of forms from the limited evidence available. This means that it is highly likely that a single occurrence of a form will come from the top-ranked variants on the curve (if we could make one), but there is also a decent chance, 20 percent, that it is actually an uncommon use somewhere in the long tail of the curve. Similarly, if we have a small number of occurrences of variant forms, it is likely, but not certain, that the distribution will begin to show a nonlinear curve. Therefore, when the amount of evidence is small, we should especially avoid the temptation to make qualitative, categorical interpretations, i.e. that the form we found necessarily was the most common one at the time, or to assume that the results are necessarily typical of the domain that we searched. The best interpretation will be one that assumes the nonlinear pattern and attempts to fit the actual returns to that pattern as well as possible. Results from small data sets will not show us the whole picture, but the 80/20 Rule offers the chance to make a reasonable estimate. And we should always recognize that, because of scaling, findings from any particular domain will not yield a result that is always generalizable to larger or smaller domains.

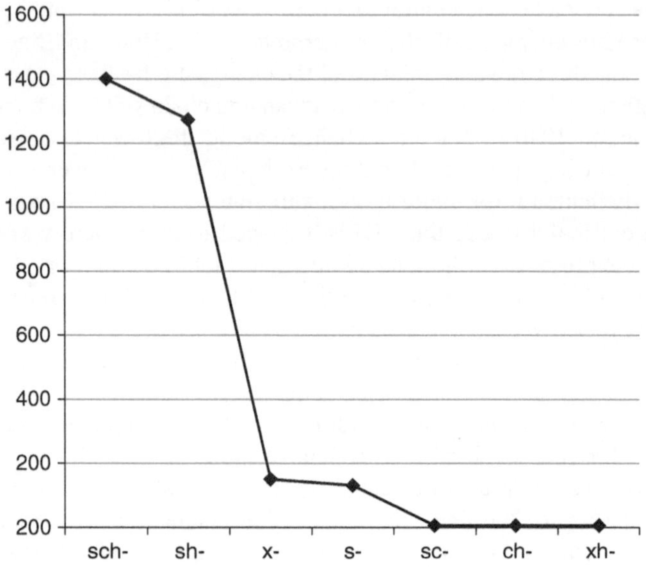

Figure 5.7 Initial consonant in *shall*, fifteenth-century Eastern texts (from Kretzschmar and Stenroos 2012: 120)

Merja Stenroos and I have offered a good example of this kind of reasoning, with application also to aggregations of features in geographical areas for Middle English dialects (Kretzschmar and Stenroos 2012), the first consonantal element of the verbs *shall* and *should*. In the Middle English Grammar Corpus (Stenroos et al. 2011), which consists of 405 text samples from the fourteenth and fifteenth centuries, eleven different spellings are found. Of these forms, only three or four occur with a reasonable frequency in any given area and period; however, the dominant forms vary greatly between areas and periods. The frequency distribution in the texts localized in the Eastern part of the country and dated to the fifteenth century (Figure 5.7) shows that the spelling <sch> is clearly the dominant one, while we know that <sh> eventually moves up to become the most frequent variant at the larger national scale in England. The most common spellings in the thirteenth century, <sc> and <s>, have, on the other hand, moved down the curve, surviving in the long tail of occasional forms found in any corpus of reasonable size.

Evidence like this of the operation of complex systems does not accord well with our traditional means of describing geographical varieties. Traditional dialect surveys in England and America have both posited the existence of dialects and used survey evidence selectively to show where they might be (Kretzschmar 2009: 66–74). Similarly, the "fit-technique" developed for

Middle English dialects by McIntosh (McIntosh et al. 1986: 23) localizes texts based on the assumption that Middle English linguistic variation forms a regular continuum into which any dialectally consistent text may be placed. Both for Middle English and for modern English, these traditional approaches apply a formal assumption of the regularity of a dialect to evidence whose variation is anything but regular. On the other hand, when we apply our knowledge of complex systems, Merja and I would prefer to map texts according to their provenance, as far as retrievable, arriving at a messier picture that simply answers the question "who wrote what where." The study of texts associated with a particular location allows us to reconstruct something of the sociolinguistic reality within the community, which may have been complex (see Stenroos and Thengs 2011). At the same time, while some scribes travelled and some reproduced the forms of non-local exemplars, we expect that the local variants will turn out to be most frequent overall in the A-curve of variant frequencies – like the one for the *shall* spellings in Eastern texts. Thus, the "fit-technique" in Middle English dialectology, like traditional modern dialectology, is best thought of as a method dominated by its formal assumptions. This does not make it bad in itself, but it does mean that, as was the case for the construction of grammars or the need to make qualitative statements about individual features, we need to understand the degree of abstraction of the "fit-technique" as it defines one dialect, one grammar, to fit all the texts of a place – because complex systems tells us that that is never actually the case in the evidence itself.

Punctuated equilibrium in language change

For one final suggestion for improvement that complex systems can bring to historical linguistics, we can pick up the geographical thread from the discussion of Middle English dialectology and combine it with the Corpus Clock. The clock is only accurate every five minutes: time slows down for a while, and then speeds to catch up. It turns out that this seemingly perverse aspect of the clock is actually a good model for what we do with long-term change in historical linguistics because it matches a model for biological evolution, called "punctuated equilibrium," that capitalizes on how complex systems work. In linguistics, many others have discussed potential parallels between biological evolution and language change, most often to reject them (e.g. Labov 2001, 2010; Lass 1990). Salikoko Mufwene has offered the latest influential parallel (2001, 2008), in which he makes the analogy of language to a "parasitic species." I would like here to say, not that evolutionary biology in the form of punctuated equilibrium provides an apt metaphor for language change as Mufwene used biology as a metaphor, but that the workings of a complex system actually account for exactly the same process in both biology and language over time.

Gould and Eldredge proposed the idea of punctuated equilibrium in 1972. As they reported in a later review article (1993), their theory by then had received wide acceptance after an initial period of misunderstanding. They did not propose, as some claimed, that species suddenly leaped to new genetic configurations (the "saltational theory"), and they had opposed attempts to coopt their theory by creationists. What they did propose was that evolution was not a slow, gradual, process that operated like clockwork; they suggested instead that in the process of evolution species enjoyed long periods of stasis during which sub-populations of the species came to coexist, whether through random genetic drift or because of differences in environmental conditions, so that change might appear to be quite sudden. Our modern view of this process is affected by what they call "the notorious imperfection of the fossil record" (1993: 222), that is, the fact that we find fossils at widely separated locations representing significantly different times, rather than collecting fossils that represent a valid sample over time of the central population of the species and of any sub-populations that may have developed. As Gould and Eldredge put it in paleontological terms, "small populations speciating away from a central mass in tens or hundreds of thousands of years, will translate in almost every geological circumstance as a punctuation on a bedding plane, not gradual change up a hill of sediment, whereas stasis should characterize the long and recoverable history of successful central populations" (1993: 222). To be clear, Gould and Eldredge are pointing out that different species will be present in the same stratum of fossils, because of the long static periods that each species can exist and because of coexistent sub-populations of the species. Before punctuated equilibrium it was thought that up to 90 percent of the differences in fossils came from the gradual evolutionary process, but now, they report, "all substantial evolutionary change must be reconceived as higher-level sorting based on differential success of certain kinds of stable species, rather than as progressive transformation within lineages" (Gould and Eldredge 1993: 223). And, to get back to the Corpus Clock, the accidents of what fossils we find can make it look like time speeds up and slows down, that change can appear to happen quickly sometimes and at other times can seem to stop. Punctuated equilibrium looks like syncopated time. The Gould and Eldredge map of the evolutionary process (Figure 5.8) contrasts the stability of the population at left (which represents gradual phylogenetic change), with the greater speciation of populations at right (which represents the simultaneous development of sub-populations). The appearance of sudden jumps in evolution comes about because imperfections in the geological fossil record may lead us to miscategorize the simultaneous existence of different populations as sequential populations. If we were to draw a straight line across the graph from any point

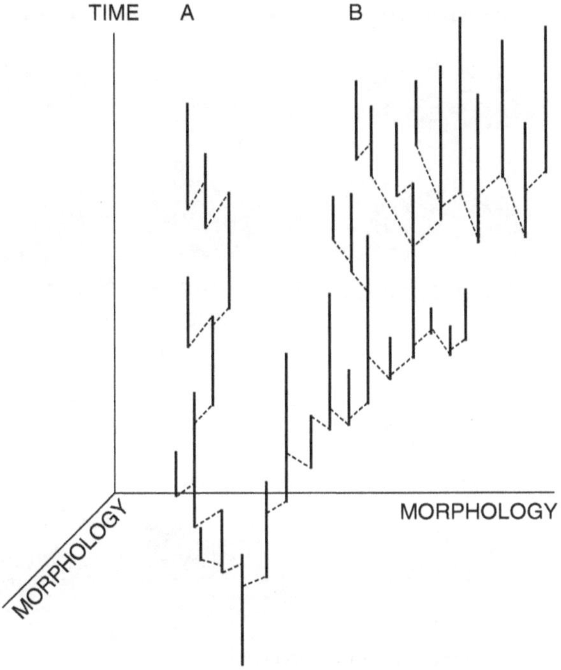

Figure 5.8 Gould and Eldredge evolutionary map (1972/1993)

on the vertical Time access (actually we should draw a plane, since this is meant to be a 3D chart), it would strike multiple populations from the sub-populations at right. Thus, from observations at any given moment in time, we are nearly certain to retrieve evidence that belongs to different populations, and the trick for us is to make an appropriate analysis of that evidence. That is, we need to deal with the appearance of slow change and fast change that emerges from such an array of facts. What must be clear, however, is that the evidence we collect from any fossil record that corresponds to the right side of the chart cannot come from a single, continuous, gradual phylogenetic process. To try to make the right side of the chart look like the left side would make us discount some of the evidence we have, or prefer some parts of the evidence over other parts, in order to force the evidence to make the neat generalization of a gradual, clockwork evolutionary process.

What Gould and Eldredge could not know in 1972, but what Gould had realized by the time he wrote his last book (2003), is that the right side of the chart corresponds exactly to what we would expect of a complex system. The scale-free property of complex systems predicts that there will always be sub-populations at different scales. We have already seen this in the distribution

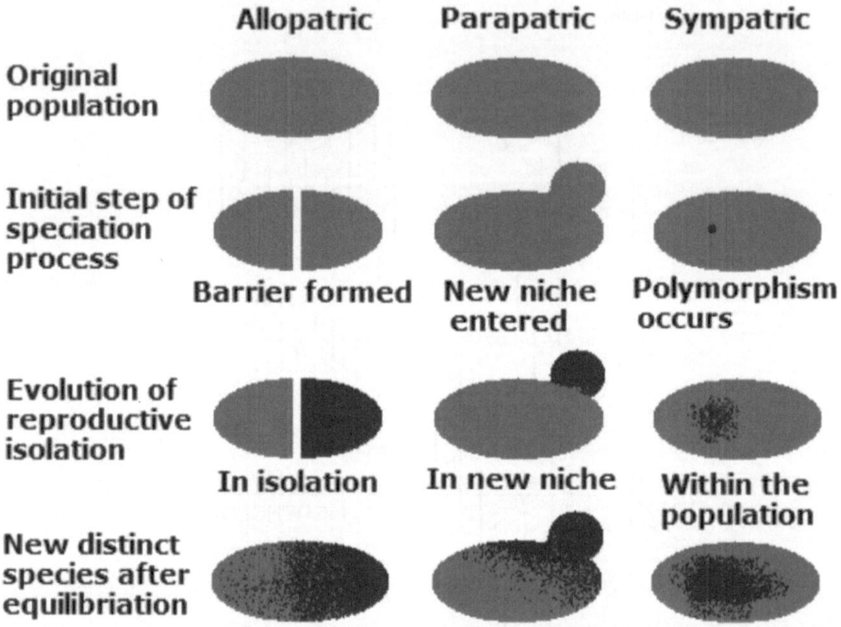

Figure 5.9 Geographic speciation modes

of individual features from the linguistic data; now we are just observing the same property at a higher level of scale of species, or in linguistic terms, in the aggregations of features we might call a dialect or a language. Gould and Eldredge originally matched their sub-populations to different geographic populations, known as allopatric, parapatric, and sympatric modes of speciation (illustrated in Figure 5.9). In each of these cases, a sub-population occurs in a geographic niche, whether created by a barrier, an extension, or an internal division. What we now understand from complex systems is that the pattern of sympatric speciation is not limited to geographic boundaries, although these remain an important consideration, but that speciation can occur for other reasons as well: the scale-free property of the complex system predicts that there will always be many sub-populations, each one defined by its own nonlinear distribution of morphological characteristics. Complex systems also tells us that it will always also be possible to consider populations at higher levels of scale, to consider any combination of sub-populations that we want to group together all the way up to the group of the population as a whole, and that at every level of scale the different morphology of the sub-populations will be expressed in the nonlinear distribution of characteristics. We see this represented at the bottom of Figure 5.9, "after equilibration," when individuals from a particular

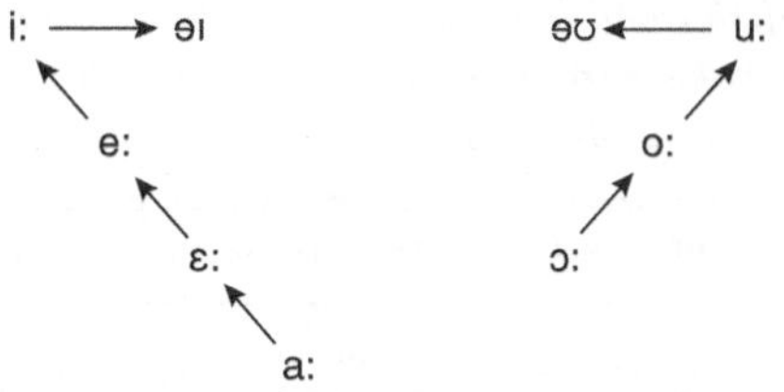

Figure 5.10 Great Vowel Shift (from glottopedia.org, under Creative Commons Attribution-ShareAlike 3.0 Unported license)

sub-population are mixed in with the population as a whole. Thus, classic examples of speciation like Darwin observed in the finches of the Galapagos do involve geography, but our new knowledge of complex systems tells us that geography is not the whole story.

An excellent example of the process of punctuated equilibrium in the history of English is the Great Vowel Shift, and Jeremy Smith has shown it to us (1996). Smith was already interested in chaos theory and complexity in 1996, but did not yet have the means to describe his evidence in those terms. Of course, now we can understand from complex systems that there will be an A-curve distribution of variants for every vowel following the 80/20 Rule, and that we can assemble the individual vowel frequency patterns to make a generalization at a larger level of scale for a regional dialect. The general outline of the Great Vowel Shift (GVS; Figure 5.10) includes the raising of both the long front and long back vowels after the time of Chaucer, roughly between 1400 and 1700. There is a long history in the literature of proposals for motivations for the GVS, often organized according to the chain-shift hypothesis that either raising of one vowel pushed the next vowel higher (a push-chain), or that raising of a high vowel created a gap in the vowel pattern that pulled the next lower vowel higher (a pull-chain). Discussion of chain shifts is still very active, and Labov 1994 (as a summary of work going back to the 1960s) relies on the chain-shift model to describe ongoing changes in American English, the Northern Cities Shift, and the Southern Shift, one step of which has already been discussed above for New York City. In an influential article, Robert Stockwell and Donka Minkova have challenged the chain-shift model for the GVS in an argument against overgeneralization of sound changes into chain shifts (1988). Their argument, that different English dialects develop their own sound changes in complex ways that make it difficult to hypothesize a language-internal general English vowel shift. Table 5.1 shows the variation in the reflexes of Middle English long, close *e* as represented in the pronunciations of a number of words by Southern speakers surveyed by the Survey of English Dialects. The evidence shows that there is substantial variation in Southern England for this vowel, and

Table 5.1 *Middle English long, close* e *class (Modern Standard English [i:,ɪi]), with speakers from the Survey of English Dialects (adapted from Stockwell and Minkova 1988)*

Form	SED 39.1	SED 39.3	Other Hampshire	Other South
fields	[i:]	[i:]	[i:, iu]	[i:, iu, iə]
feed	[i:]	[e:]	[i:, e:, ai]	[i:, ɪi, ɪ]
keep	[ɪ]	[ɪ]	[ɪ, i:, ɪi]	[ɪ, ɪi, i:]
cheese	[e:]	[e:]	[i:, ɪi]	[i:, ɪi]
yeast	[i:]	[ai]	[i:, ɪi]	[i:, ɪi, ɛ]
niece	[e:]	[ai]	[i:, ɪi, ei, e:]	[i:, ɪi, ei, e:]

that the variants represent different stages of the GVS. Stockwell and Minkova conclude that (1988: 378):

In the so-called chain shift it is more likely that schoolmasters and language-conscious educators made choices in codifying the standard, choices that favored pronunciations that they believed, rightly or wrongly, resulted in greater clarity of expression. These choices resulted in a system which, viewed from the narrow perspective of any version of purely structural linguistics, appeared to form a neat cohesive sequence of events within a homogeneous dialect community which in fact surely never existed.

Our knowledge of complex systems allows us to go one step further and accept the frequency profiles from individual speech behavior within an area that form A-curves, in addition to talking about the complexity and fluidity of dialects across the areas of a region. Unpredictable elements in individual speech do not necessarily mean that these speakers are in any way language impaired.[3] Some of these unpredictable characteristics, whether the result of creativity or chance, whether above or below the level of personal or social consciousness, remain on the A-curve, and many of the variants in the long tail either maintain variants that were more common in the past, or represent variants that may become more common in the future.

As we have seen in the discussion above of Labov (1994), the most spirited defense of language-internal chain shifts, highly subjective and selective use of the dialect data can generate arguments for or against particular interpretations of the development of sound changes as chain shifts. The overall findings of survey research, however, remain at odds with Neogrammarian ways of thinking about sound change in historical linguistics. Besides a mechanical

3 It has been suggested that an inability to manipulate the linguistic variation in a community is actually a pathological condition. Perhaps the best literary example of the point occurs in John Kennedy Toile's *A Confederacy of Dunces* (1980), in which the main character is handicapped by his inability to shift out of his Standard English speech in order to deal effectively with the many different dialect speakers in the New Orleans setting of the novel.

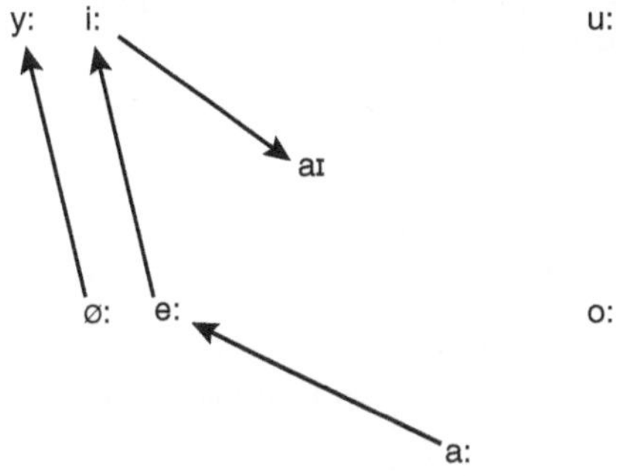

Figure 5.11 Great Vowel Shift, Northern region (adapted from Smith 1996: 100)

understanding of sound laws and language change, the Neogrammarians also promoted the use of scientific methodology and close attention to actual data.[4] The Neogrammarians can also offer an emphasis on individual speakers and their speech habits or Language Custom (*Sprachusus*; see e.g. Herman Paul (1889), whose ideas received extensive comment in Weinreich, Labov, and Herzog (1968). Indeed, the theory of mechanical sound change might be thought to impose an impossible burden on this part of the Neogrammarian program, since mechanical sound change introduces the methodological requirement of a relatively fixed notion of speech communities, when complex systems instead offer us evidence for scale-free networks.

To return now to Jeremy Smith, he develops the idea offered in Stockwell and Minkova (1988) that regions could have their own distinctive changes that, together, we might recognize as the GVS. He points out that in the North, the Great Vowel Shift did not proceed according to the usual overall generalization: an extra front vowel was raised, and the back vowels remained unraised (Figure 5.11). Smith offers us complications of the Great Vowel Shift in the South, too (Figure 5.12). He describes two competing systems in fifteenth-century London, the first among those whom he calls "descendants of Chaucer," from higher

[4] Robert B. Lees took the opposite point of view in his highly approving review of Chomsky's *Syntactic Structures*:

> It is a common misconception on the part of many a scientist, strange to say, that correct scientific theories are discovered by making many observations of nature, that somehow the right answers just leap up out of the laboratory notebook if only we have measured enough things accurately . . . the theories by means of which we order our experiences, on the street or in the laboratory, are generated only by those flashes of insight, those perceptions of pattern, which mark off the brilliant scientist from the dull cataloguer of data. (1957: 379–380)

It would be hard to imagine a more romantic notion of science than this one, a notion of science more dependent upon intuition and insight.

Table 5.2 *Three systems involved in the GVS, ca. 1650 (adapted from Smith 1996: 108–110)*

System I:	
Middle English	/e:/ > /i:/, e.g. *meed*
	/ɛ:/ > /e:/, e.g. *mead*
	/a:/, /ai/ > /ɛ:/, e.g. *made, maid*
System II:	
Middle English	/e:/ > /i:/, e.g. *meed*
	/ɛ:/, /a:/, /ai/ > /e:/, e.g. *mead, made, maid*
System III:	
Middle English	/e:/, /ɛ:/ > /i:/, e.g. *meed, mead*
	/a:/, /ai/ > /e:/, e.g. *made, maid*

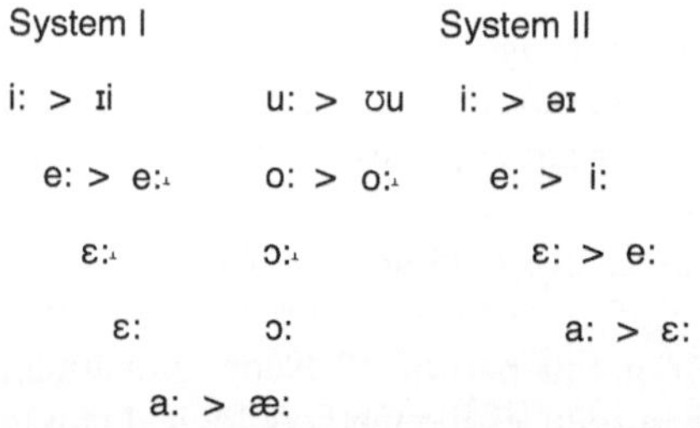

Figure 5.12 Great Vowel Shift, London (I) and the Midlands/South region (II) (adapted from Smith 1996: 102)

social circles, and the second system prevailing among the remaining speakers in the Midlands and South. System I in London has only subtle raising in the early fifteenth century, while System II in the Midlands and South has full-step raising of vowels. We see here geographical sub-populations, but also a pair of social sub-populations within a region. Somewhat later in his argument, Smith also introduces a third system which originated in East Anglia but whose speakers also went to London, where their speech pattern competed with that of the descendants of Chaucer; speakers retained System II longer in the countryside (Table 5.2).

This amounts to a kind of linguistic equilibration, as described earlier for modes of speciation. Smith provides orthoepic evidence and rhymes to support his argument, which are not reproduced here. The point of his argument for our purposes is that complex systems predicts that both the regional and social sub-populations could develop different characteristics, and that these sub-populations could coexist at the same time, and that, just as in punctuated equilibrium, we can observe "equilibration" as members from one sub-population become mixed with others, as Smith claims that speakers did mix in London. Of course, each of the sound changes suggested by Smith ought to be supported, if we had enough evidence, by a sequence of A-curves that showed the change in frequency of each realization over time. So, the linguistic evidence is not just parallel to the fossil record. It is the same procedure to use the fossil record to describe species as it is to use fragmentary linguistic evidence to describe competing dialects in English – or in the longest historical perspective, competing languages in the Indo-European family. The same underlying process is at work for both paleontology and historical linguistics. To rephrase Gould and Eldredge, while we used to worry about dialect contamination in the gradual evolution of a language, now all substantial evolutionary linguistic change ought to be reconceived as higher-level sorting based on differential success of certain kinds of stable language varieties. While the kind of sorting carried out by the Neogrammarians in the nineteenth century to describe the daughter languages of Indo-European was the best detective work of its age, based as it was on qualitative statements about features and words, we can now accept the idea of punctuations, scale-free networks of speakers, and do a better job of matching populations and sub-populations to our evidence.

Given Smith's evidence and my argument, then, what should we make of the GVS, and of other historical accounts of major sound changes? The complex systems approach would suggest that we should not abandon them. The GVS remains a useful generalization at the top level of scale. But we cannot just work at the top level of scale, because we are then subject to the dangers of misinterpretation of our imperfect evidence that come from punctuated equilibrium. Just as for the fit-technique, we cannot apply the top-level generalization of the GVS back down onto primary evidence, because it is based on the formal abstraction that we can define one grammar to fit all the texts of a place – and complex systems tells us that that is never actually the case. For changes lower down the scale from the GVS, we should attempt to make better statements that connect the evidence for a change with the generalization. So, for instance, Moore and Knott offer a reference grammar with no fewer than sixty-eight paragraphs describing sound changes between Indo-European and Late West Saxon (1972). Especially for the paragraphs on the later changes for which manuscript evidence exists, it would be useful to replace adjectives like "sometimes, often, frequently" with an account of the evidence that might show in which sub-populations we can expect higher frequencies of the changed forms.

Just as we must do with the Cambridge Corpus Clock, we have to expect that time in some places will appear to go faster and in some places slower in historical linguistics. We now know that linguistic change does not just run like clockwork. There were good reasons why the Neogrammarians created a mechanical process for linguistic change – chiefly to adopt the best science then available to govern their work. We, too, need to do this: we should not just accept traditional methods because they are traditional; we should adopt the best science available now to govern our work, and complex systems now serves that purpose. If we choose to make formal generalizations, and in historical linguistics a great many scholars do, then any formal statements we make should reflect what we now know about the emergence in our evidence of scaling, nonlinear frequency patterns from the complex system of speech.

6 Neural networks and complex systems

Aesthetics and perception are inextricably bound together, and not in a simple way.[1] As we have seen in previous chapters, our perception of language around us is an essential component of the operation of complex systems because that is how we share information about language, and then build nonlinear frequency distributions. Aesthetics can add a social component to our perception as individuals. As the *OED* tells us in its etymology for *aesthetic* (*s.v.*, a little gem of an etymological essay) the word comes to us from Greek, where it had to do with our perception of material things, as opposed to our understanding things of the mind. This is the sense in which the word *aesthetic* is used by Kant. But spreading from eighteenth-century Germany to other places and languages, another meaning for the word became popular, the sense 'taste.' *Taste* can be either "good" or "bad," and so in this sense we see that perception is conditioned by social convention. The *aesthetics* of an object or situation thus, in popular usage, comes to refer to something like 'conformance with accepted canons of beauty,' or even 'conformance with social conventions.' Today, then, we have two sharply different current meanings, the neutral sense of 'sensual perception' in the philosophical field of *aesthetics*, and a more popular meaning in which perception receives cultural shaping into 'taste.' C. S. Lewis (1960) might say that this popular meaning has become the "dangerous sense," the one that we think of first when we encounter the word, even though it may actually mean something else in context. It will probably not cause problems if we think that *aesthetics* always refers to social conventions, for it is not dangerous, though perhaps a bit jaded, to elevate social conventions in that way. However, we are at risk of being seriously misled if, when we talk or hear about *aesthetics*, we think that the word merely refers to 'perception' when in fact the perception at issue may not be purely objective but instead is actually guided by social convention. In such a case we risk losing philosophical objectivity or our scientific viewpoint, and replacing it with 'taste.' On the other hand, when we think that we merely perceive something, we may actually behave aesthetically, by invoking social conventions as we receive and interpret information.

[1] This chapter is based on Kretzschmar (2008).

We always need to be clear whether we are talking about 'taste' or about philosophy and science.

This bald distinction between 'taste' and perceptual aesthetics is not meant to minimize or trivialize serious philosophical and academic treatments of aesthetics. For example, H. R. Jauss has offered a fine historical overview of aesthetics in relation to literary study that begins with Plato and Aristotle and ends with modern notions of aesthetics from the formalists and Marxists (1982: 3–151). The distinction is just meant to highlight the distance in everyday language between meanings of *aesthetics* and *perception* that, without reflection, can cause misunderstandings in practice. The word *objective* is similarly bound up with perception. What we may take to be objective observations, statements of fact not colored by merely personal or conventional judgment, are actually culturally dependent. The famous Sapir–Whorf Hypothesis about relationships between language and culture tells us so (see Carroll (1956) for several fascinating essays by Whorf), and the culturally contingent nature of common categories like colors, where terms are matched in different ways to the continuum of frequencies of light, has been telling us the same thing in a different way beginning with Berlin and Kay (1969). Still, as with the distinction regarding *aesthetics*, it will be useful to employ the term *objective* here, not to indicate a strong position regarding the reality as fact of scientific observations, but to enact the distinction in everyday language between such "scientific" observations and those of a more subjective nature, meaning observations either intentionally influenced by personal views or observations which may be colored by conventional judgments with or without the observer being conscious of the influence.

The traditional notion of a language or a dialect offers a good example of how aesthetic judgments can be bound up with what we believe to be objective observations. One may characterize the "linguistics of linguistic structure" that has dominated linguistics since Saussure as making the assumption that languages are "systems," entities that belong not to individuals but to the collectivity of the community. In Noam Chomsky's famous words, such systems entail the assumption of an "ideal speaker-listener, in a completely homogeneous speech community" (1965: 3). Modern sociolinguistics shares this assumption; as Shana Poplack put it (1993: 258), sociolinguists try "to gain access to the vernacular, the relatively homogeneous, spontaneous speech reserved for intimate or casual situations. This is taken to reflect the most systematic form of the language acquired by the speaker ... " These ideas, however, arise from our perceptions of language around us, not directly from the language around us. When we notice that one variant of some linguistic feature appears to be the most frequent one for a group of speakers, we commonly make the generalization that the variant is "normal" for everybody in that group, and that everybody habitually uses it. The group may be at the level that we associate with languages (one group of speakers who say *vous*, in French, as opposed to another group of speakers who say *you*, in English), or may be at the level that we associate with dialects (one group of speakers who say *y'all*, in the Southern American English dialect, as opposed to the speakers who say *you*, in the Northern American English dialect). Taking many features together at once, we make the further generalization that a language or a dialect exists, composed of the normal use of some set of particular features by its speakers. We may think that American Southerners, for instance, must all say *y'all* for *you* plural, say *far* and *tar* for *fire* and *tire*, and talk slowly with a drawl. This kind of generalization is an observational artifact (in the term first used in Günther et al. 1996) because it always misrepresents the actual distribution of variants in the group – which in fact, as we now know, always has an A-curve pattern. *Y'all* is a very frequent variant for *you* plural but not everybody uses it and those who do will not use it all the time, and there is actually lots of variation in how people pronounce *far* and *tar* for *fire* and *tire*. As Southerners are aware, and the rest of us, too, if we really thought about it, there are many flavors of Southern speech (see Kretzschmar (2008) for a full account). Our perception of the frequency profile of variants represents the first, objective meaning of aesthetics, while the traditional interpretation of languages and dialects from Chomsky and Poplack represents the second, socially conventional sense of that term.

Non-linguistic information also very often creeps in to affect our perceptions of language. The situation of so-called "Standard" varieties of languages provides a good example of how perceptions thus can distort the actual facts of production of linguistic features. The notion "Standard," as distinct from "normal," represents institutional agreement about which variants to prefer, and enacts our perception that some variants are better somehow than other variants. In terms of aesthetics, "Standard" is thus a matter of taste, not of objective observation. Some of the perceived tasteful variants are far less frequent than the "normal" variants for many groups of speakers – how many people would actually say "It is I" or know about and act upon old grammatical shibboleths like *shall/will* or *who/whom* or split infinitives? This creates the perception of parallel systems for "normal" speech in any number of groups vs. a monolithic "Standard." There is no population of speakers in America (or elsewhere) who speak Standard English as a native dialect, not Midwesterners, not people from Michigan, even though these places are often named as locations where Standard American English comes from. The "Midwest" is a handy term for this purpose, since different people would include different states as belonging to it (Minnesota? Kansas? Kentucky?). Michigan speakers in fact often participate in the "Northern Cities Shift," a label for ongoing change in pronunciation of vowels proposed by Bill Labov (see Labov et al. 2006), which makes them sound distinctive to speakers from other regions, not "Standard" as the Michiganders themselves often think. The situation for Received Pronunciation (RP) in Britain is parallel, but more complicated socially (Clive Upton's representation of British English with a more "northern" flavor in Upton, Kretzschmar, and Konopka (2001) has come in for substantial comment). Our perceptions are thus guided by the A-curve distributional patterns of language production, but perception is not the same as production: we form impressions of speech that may overgeneralize frequent features into categorical features, or impressions of speech in which our aesthetic judgments, in the second sense, may lead us away from actual speech as people use it and towards conventional interpretations such as "Standard." In sum, then, traditional ideas of dialects, including the "Standard," arise from the second definition of aesthetics; they are conventional entities rather than anything that arises directly from observations of language in use.

The argument in this chapter partly concerns technical matters, but its larger point parallels the two meanings of *aesthetics*, to try to separate objective perceptions from traditional conventions, with reference to neuroscience and neural network simulations based on it.

Models in cognitive science

Contemporary cognitive science explores the ways that information, our perceptions in the original sense of the aesthetics of material things, can be processed and stored in the brain. Two main theories of cognitive information processing have dominated the development of the field. Earlier ideas of cognition relied on a behavioral model, but beginning in the 1960s and 1970s two new models arose: a formal model for physical symbol systems based on Chomsky's *Syntactic Structures* (1957), and a computational model for neural networks based on algorithms and, as developed by Alan Turing, a "definite procedure" for the steps to process information (on Turing and definite procedures, see Mitchell 2009: ch. 4). Bermudez describes these two models (2010: 457):

> According to the physical symbol system hypothesis, we need to think about information processing in terms of the rule-governed transformation of physical structures. These physical structures are information-carrying representations. Neural network modelers think of information processing somewhat differently. Information in neural networks does not have to be carried by discrete and independent structures. It can be distributed across weights and connectivity in a neural network. And information processing seems to work differently. The algorithms that update neural networks and allow them to learn are very different from the rules invoked by the physical system hypothesis.

The key difference between the two models is the notion of a physical structure, a representation, that bears information. Neural networks can carry information without representations, among the constant firing of interconnected neurons.

As we shall see, neurolinguistics and cognitive linguistics have traditionally relied on representations. Neuroscience more generally, however, has moved towards neural networks and dynamic processing, towards complex systems. Bermudez takes a balanced perspective between the models, suggesting that a hybrid mental architecture may be possible, and that each model may be suited to different information-processing situations.

As originally defined by Newell and Simon in 1975 (see Bermudez 2010: 146–154), physical symbol systems rely on four basic ideas:

1 Symbols are physical patterns.
2 These symbols can be combined to form complex symbol structures.
3 The physical symbol system contains processes for manipulating complex symbol structures.
4 The processes for generating and transforming complex symbol structures can themselves be represented by symbols and symbol structures within the system.

According to this model, then, thinking consists of transformations of symbol structures according to rules. Structures of symbols consist of the physical relationship of the symbols. The processes that operate on the symbols and their structures are characterized by Newell and Simon as "processes of creation, modification, reproduction, and destruction" (Bermudez 2010: 147). The development of physical symbol systems that tends towards language is Jerry Fodor's language of thought hypothesis (Bermudez 2010: 156–164). Fodor modified the hypothesis to include three key points (Bermudez 2010: 160):

1 Causation through content is ultimately a matter of causal interactions between physical states.
2 These physical states have the structure of sentences and their sentence-like structure determines how they are made up and how they interact with each other.
3 Causal transitions between sentences in the language of thought respect the rational relations between the contents of those sentences in the language of thought.

Thus for Fodor the physical proximity of physical systems has been turned into a logical language, in which there is both a semantics of the symbols and a syntax of their organization into sentences. And thus, too, the natural relationship between this branch of cognitive science and linguistics. Neurolinguistics follows from the physical symbol system hypothesis and the language of thought. In Bybee (2001), for instance, it is proposed that the unit of storage for phonological information is the word. As we have seen, usage-based linguists regularly talk about cognitive structures in terms of representations. Exemplar theory and usage-based linguists refer to prototypes, a cognitive model that applies to serial processing of information. We may ask ourselves, however, how much of neurolinguistics is a chicken-and-egg problem, where linguists' expectations of language as from Chomsky and the generative tradition influence their thinking about the cognitive status of language. Linguists' expectations are thus understandable as a case of aesthetics in its second sense, one in which our perceptions are shaped by social environment.

The neural network alternative will be presented below in an extended experiment. Its focus is the relatively unknown part of cognitive functions between large-scale brain functions and the electro-chemical function of individual neurons. Contemporary neuroscience offers a physiological description of the operation of language. Broca's and Wernicke's discoveries in the nineteenth century of brain regions related to language were followed by study of many aspects of aphasia in the twentieth century. Such study was primarily based on patients with brain damage and consequent speech impairment. Neuroscience has recently enjoyed greatly expanded possibilities owing to the invention of non-invasive instrumental means to study brain function, notably PET and fMRI scanning and the use of EEG. These methods can be used to study normal subjects, instead of having to predict normal function primarily from patients with impaired abilities. At the other end of the spectrum, the function of individual neurons can be studied (though not in humans) by using microelectrodes to study electrical activity and activation levels over time. Between these two poles, however, lies the connectivity between neurons, and as Bermudez suggests, "The most fundamental feature of the brain is its *connectivity* and the crucial question in understanding the brain is how distributed patterns of activation across populations of neurons can give rise to perception, memory, sensori-motor control, and high-level cognition" (2010: 217). Computational neuroscience and connectionists employ neural network models in order to understand, say, the process of language acquisition among children. In the experiment presented here, evidence from survey research will be submitted to neural network algorithms in order to understand how a speaker might perceive and process such evidence over time.

According to this model speech in the brain is not a single process but a massively interconnected one, a neural network. According to the prevailing connectionist model (e.g. Edelman 1987; Pulvermüller 2003), neurons develop billions of connections in a massively parallel network, in which no action or perception could be considered to have a single or simple "impulse," as earlier theory suggested. Bybee's proposal that phonological information is stored on the basis of words (2001) must be understood in neuroscience not as the brain having some single physical location to store a word, not as a representation, but rather as the brain having a collection of interconnected neuronal pathways whose activation is related to a word. In hearing, for instance, Saussure's (1916 (1986)) seemingly unitary "impulse" corresponds to thousands of neural fibers (specialized hair cells) in the inner ear that vibrate in response to specific frequencies, and in turn activate neurons that fire repeatedly and transmit signals to the brain. Thus the brain receives patterns in response to a speech sound, changing every moment, that might be compared to the pixels on a visual monitor, some turned off and some turned on at any given moment in response to the stimulus of physical sound waves; just as what we consider to be a picture on a monitor actually consists of a configuration of pixels, we can recognize a "speech sound" as a particular configuration of neurons firing in response to input from the neural fibers. Frequent patterns habituate neuronal pathways as we "learn" speech sounds just as, to continue the visual metaphor, some pixels could become burned in on older monitors (thus the need for screen savers to prevent what was considered to be damage, not learning). The emergent order of the network is what actually constitutes what we consider to be a unitary speech sound.

The physical symbol structure and the neural network models have more recently been supplemented by a dynamical systems model that does not rely on either rules or definite procedures. Dynamical systems rely on coupled elements and feedback that work together, while the system remains in motion, to create a probabilistic profile, called an "attractor," out of the state space of possible outcomes. Dynamical systems thus replace sequences of actions determined by either rules acting on representations, or algorithms acting on information with definite procedures, with a kind of juggling act where the information available mutually influences itself. We will thus be interested not in some state of the system after the application of a rule or the evaluation of algorithms, but in the emergent pattern of the state space as it continues to change over time.

A good example of the operation of a dynamical system is the governor invented by James Watt for the new technology of the steam engine, based on mechanisms already in use in windmills (as reported in Bermudez (2010: 417–422), following Van Gelder). Watt had invented a way to use steam engines to produce rotational movement using a flywheel, and he needed a way to manage two variables to keep the speed of the flywheel as stable as possible, steam pressure and the load from work being done. While it might be possible to compute the pressure and load and thus to adjust the throttle as needed, this method can only make adjustments on the schedule for making the computations. Watt, on the other hand, created a method for continuous adjustment. Two metal balls attached to the flywheel rise or fall according to centrifugal force as the speed of the flywheel changes, and the throttle is attached to the mechanism of the metal balls so that it changes in inverse proportion to changes in speed of the flywheel. If the load increases and the balls fall, the throttle increases; if the load decreases and the balls rise, the throttle decreases. The Watt governor thus operates continuously without computations made at different states of the system. Its central features are the interactions between the components and timing, as the governor remains in motion. No rules as from physical system structures or algorithms applied to states as from neural networks are involved. In cognitive terms, if the Watt governor model applies to brain function, the interaction of connected neurons and the timing of the firing of neurons, multiplied by all of the things we think about, are sufficient to account for management of information without representations and rules to manipulate them, and without algorithms applied to states.

Dynamical systems, as should become obvious, is another name for the operation of complex systems as described in Chapter 1 and throughout this volume. Neither neural networks nor dynamical systems employ representations, and so they might be considered to be more similar to each other than they are to the physical symbol systems model. What separates them is that neural networks pass through a great many interactions to reach an outcome, while in dynamical systems the process is more important than the particular state of the system at any one time.[2]

Neural network simulations

The network simulations presented in this chapter, of course, are not the same as making instrumental observations of brain function. Some might say that neural network algorithms may well be good examples of the science of statistics, but might wonder how such simulations are relevant to neuroscience itself. In the current state of the field of neuroscience as it relates to language, instrumental observation of brain function through fMRI or other techniques is usually carried out for only one word at a time as a linguistic stimulus (Xu et al. 2005), however, does consider longer stretches of language). If we are to answer questions about cognitive processing of a great many words, including how we make decisions between alternative words about what to say or what to understand, which we certainly want to do, then we must therefore simulate brain function through a neural network algorithm. The scale of the experiments reported here, with data in the aggregate from over 1,000 speakers, at least gives us a chance to observe strong effects, if not finely graded information of the kind that we might get from instrumental observation of the brain itself. While the simulation yields only indirect clues about cognitive function, such clues are the best we can hope for at this moment in time if we are to ask the larger questions. And with indirect clues, when we have to fill in the inevitable gaps in our picture of cognition, it is all the more important to consider aesthetic issues, to separate conventions from objective perceptions.

[2] Lakoff and Johnson (1980) and Lakoff (1987) are essentially connectionist arguments about language, but the assertions there about mental schemas branch across the models by claiming special status for elemental metaphors that look like representations.

The neural network experiments are part of a series of papers in which I have discussed the application of advanced statistical methods to language data from survey research (e.g., Kretzschmar 1996, 2003, 2006, 2011a); Kretzschmar and Tamasi 2003). The overall point of these papers has been to show that we cannot expect our common knowledge about language to be supported by the actual facts of language variation as shown by survey research, especially our universal belief in the real existence of dialects as subsystems of languages. The possibility for intensive computational analysis of large data sets, I have argued, should make us modify our traditional assumptions about language in use, in favor of new ideas that better reflect the data that we collect. No matter how advanced our statistics and our technical capabilities, they will not be the most useful as long as we do not match our new methods to appropriate characterizations of the evidence, in this case to what people actually say.[3] In order to address issues in neuroscience, we will explore the relationship between the results of computational analysis of language data with neural network algorithms, traditionally accepted dialect areas and groupings, and a new characterization of order as it emerges from speech interactions.

Neural network analysis as applied here to speech evidence is modeled on how learning occurs in the brain. A compact set of Boolean rules controls the network, just as the firing of neurons controls brain function. As Bermudez describes the model (2010: 234–235):

> Information enters the network via the input layer [Figure 6.1]. Each unit in the input layer receives a certain degree of activation, which we can represent numerically. Each unit in the input layer is connected to each unit in the next layer. Each connection has a weight, again representable numerically . . . The activation at a given input unit is transmitted to all of the units to which it is connected in the next layer. The exact quantity of activation transmitted by each unit in the input layer depends upon the weight of the connection. The total input to a given unit in the first hidden layer . . . is the sum of all the quantities of activation that reach it. If the total input to the unit reaches the threshold then the unit fires (i.e. transmits its own activation). The amount of activation that each unit transmits is given by its activation function . . . In the figure the activation function is a sigmoid function. This means that a_i is low when total input (i) is below the threshold. Once the threshold is reached, a_i increases more or less proportionally to total input. It then levels out once the unit's ceiling is reached.

Input information is not automatically carried through the network, but is modified by weights as it is transmitted through a "hidden" layer, which is neither the input nor the output layer but something in between. So, in a hypothetical

[3] To bring up a parallel from physics in the last century, no matter how good their techniques of measurement became, they still did not find and characterize the phlogiston or the ether that physicists at the time thought there had to exist. The neural network algorithm was implemented by Jean-Claude Thill, without whose collaboration these experiments would not have been possible.

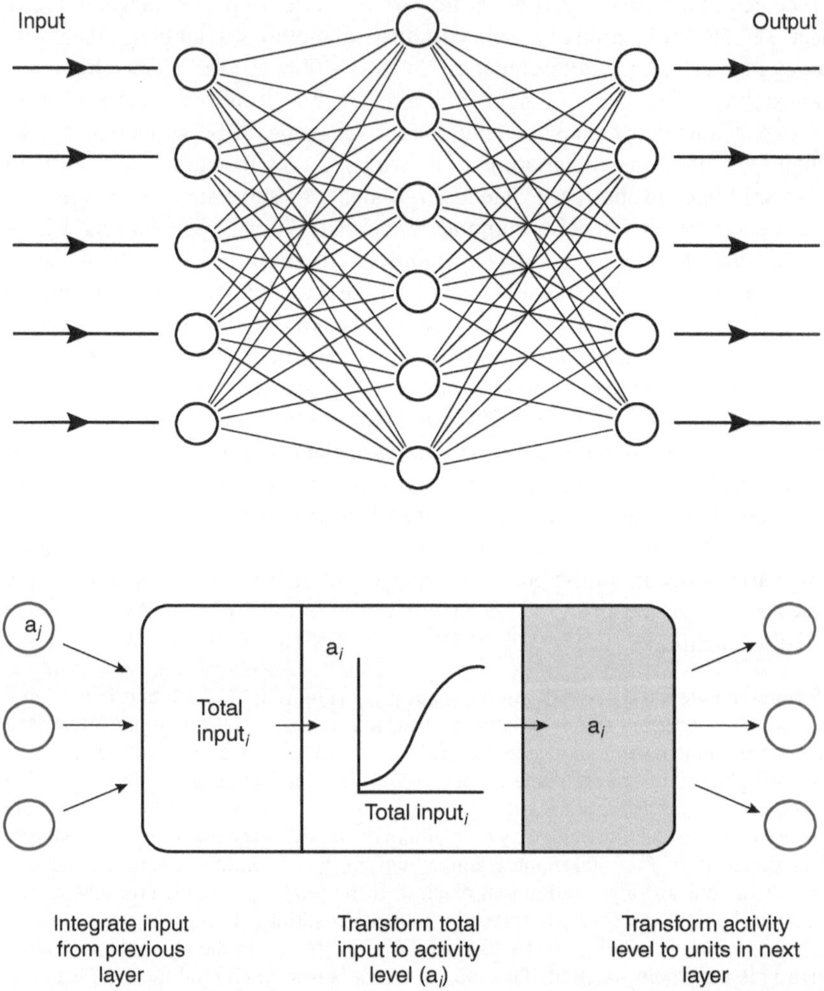

Figure 6.1 Neural network processing model (adapted from Bermudez 2010: 234)

example, sensory input received from the ears might be weighted heavily at nodes in the hidden layer that process sound, but would not be entirely separated from tactile information (moderately weighted because sound consists of vibration) or visual information (with low weight). Activation amounts, too, are variable, sensitive to an activation threshold and thereafter to additive increases with more input. The learning process thus does not work directly and mechanically from input but involves weights and thresholds that are subject to change over time. In this view, learning necessarily involves management of information, not just storage of information.

When the neural network process is modeled with data and algorithms on computers, learning requires thousands of iterations in which the same data is reprocessed many times so that the connections and weights self-organize. The process begins with a seed, a random set of groups, and after each iteration of the algorithms the groups are adjusted so that some data elements move into different groups and the weights for different data elements may also change. After thousands of repetitions, the data will have been sorted into an array of groups, each of which contains some of the data at specific weights. The analyst may then observe what groups have been created and what weights have been applied as the result of the entire run of the neural network.

Self-Organizing Maps

The technique presented here, Self-Organizing Maps, reduces the dimensionality of variation in complex data in order to find the most likely patterns in the data, or "nodes" (see Skupin and Agarwal (2008) for general discussion; see Kretzschmar (2011a); Thill, Kretzschmar, Casas, and Yao (2008), for particular language applications). A node for the Atlas survey data is a particular pattern of the words from the domain, such that, for some group of speakers out of the 1,162 LAMSAS speakers, each word in the domain tended to be present or tended to be absent with a specific weight determined by the iterations of the process. For example, the *Hearth* words discussed here (listed in Table 6.1) include twenty-two different fairly common response types from four different survey questions in LAMSAS data. Any given speaker will have a particular pattern of response types – in responding to the survey, the speaker will have used some of the words on the list, and not used others. Thus, for a single speaker, a pattern would consist of something like "*andiron* yes, *dog irons* no . . . *lightwood* yes, *pine* no, *pitch-pine* no, *rich-pine* yes." Within a group of speakers, any of the words may have been used by all of the speakers, by some subset of the speakers, or by none of the speakers. A possible node, then, consists of the pattern of weighted occurrence of the response types for some group of speakers, like ".396 *andiron*, .263 *dog irons* . . . .209 *lightwood*, .143 *pine*, .288 *pitch-pine*, .288 *rich-pine*." For a set of twenty-two items, each of which may be present or absent, there is a state space equal to 2^{22} (4,194,304) possible patterns for any single speaker; this figure rises by several orders of magnitude when percentage weights are allowed. More orders of magnitude come from study of a group of speakers that may include anywhere from one individual to all 1,162 speakers in the survey. All of these possibilities come from the common lexical variants of just four everyday objects, which is complex enough for an experiment but very far from the myriad dimensions of

Table 6.1 Hearth *words*

Common variants for object that holds up logs in a fireplace	Common variants for the log at the back
andirons	backlog
dog irons	back-stick
dogs	chunk
firedogs	log
fire-irons	
handirons	
Common variants for the shelf over the hearth	**Common variants for the material used to start the fire**
fireboard	kindling
mantelboard	lighterd
mantelpiece	lightwood
mantelshelf	pine
mantel	pitch-pine
shelf	rich-pine

variation in the words, pronunciation, and grammar of language as we use it. Out of billions of possibilities for this experiment, then, the neural network algorithm will select the 100 "best" nodes. It is hard to know what "best" really means here, but the process does always classify all 1,162 speakers in its nodes.

As Jean-Claude Thill has implemented the Self-Organizing Map algorithm for this experiment,[4] the selection process starts with a set of "seed" nodes randomly set out in the 10×10 matrix, and then each actual speaker pattern from the survey is matched to the node whose pattern fits it the best. As part of the matching process, the weight assigned to each word (as a dimension of variation) in the node pattern is changed slightly, and the weights of words in neighboring nodes are also adjusted to a smaller extent, as each speaker is associated with the node. Then, the matching of speakers to nodes is carried out all over again, and the same speaker may well be matched to a different node because of the adjustment of the weights. During this next iteration of matching, the weights are all adjusted yet again to go along with the new assignment of speakers to nodes. And this process is repeated again and again. Eventually, the nodes will settle down into a stable, homeostatic state of order,

[4] Our work on implementation and testing of neural network analysis was supported by NSF SBR-9975657, "Collaborative Research on the Geography of English Dialect Features by Self-Organizing Maps." The choice of a 10×10 matrix of 100 nodes is essentially arbitrary. We have also used a 5×5 matrix, with similar results. As a practical matter, it has turned out to be useful to have 100 nodes in the matrix rather than a smaller number, because the larger number of nodes allows us to see more clearly the effects presented below.

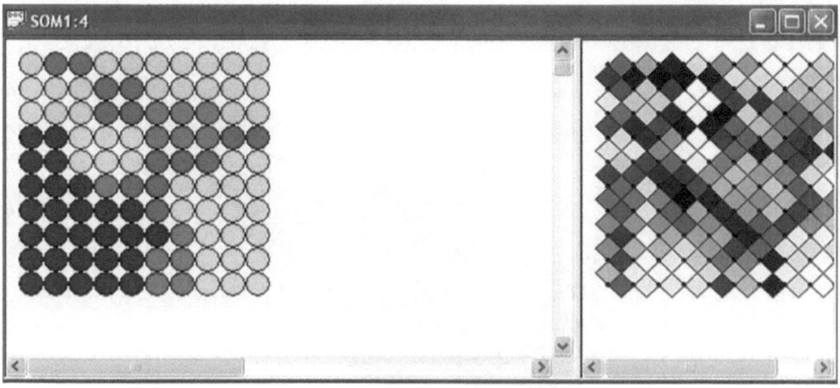

Figure 6.2 Matrix of nodes (10×10) in colored patterns

and the analysis is complete. For an experiment of the size indicated here for *hearth* words, about 25,000 iterations are required to achieve the stable state. When it does occur, the node weights are established, and the nodes themselves also turn out to have arranged themselves so that similar nodes are usually near each other in the 10×10 matrix. This makes it possible for the algorithm to select from two to ten sets of patterns, each represented by a different color, into which the nodes themselves might be classified based on the relative similarity of the patterns they represent, illustrated in grayscale at left in Figure 6.2. A separate display at right shows the degree of relationship between neighboring nodes, with darker shading to represent greater closeness of the nodes. When the sets are complete, the user can inspect the weights in the nodes, and can plot geographic locations of the speakers in any node or combination of nodes.

Unfortunately, results to date have seemed disappointing when applied to geographic language variation (Kretzschmar 2006; Nerbonne and Heeringa 2001). Neural network analysis cannot be shown reliably to replicate traditional dialect patterns when the speakers in the node patterns are plotted on a geographical map. Instead, self-organizational patterns yielded by neural net algorithms appear to respond only in a general way to assumed dialect areas. Figure 6.3 highlights the speakers from a node pattern that may resemble the Midland and Southern pattern from Hans Kurath's famous 1949 description of American dialects, based on the same data. Node patterns often appear to be derived not from speech per se but from conditions of data acquisition such as "fieldworker" effects (Kretzschmar 2011a). Figure 6.4 shows highlighted speakers whose commonality surely comes from the fact that most of them were interviewed by Raven McDavid, who conducted the interviews in Upstate New York and in the far south. Many node patterns yield scatterings of speakers (Figure 6.5), or rough groupings with many outliers, as in Figure 6.6,

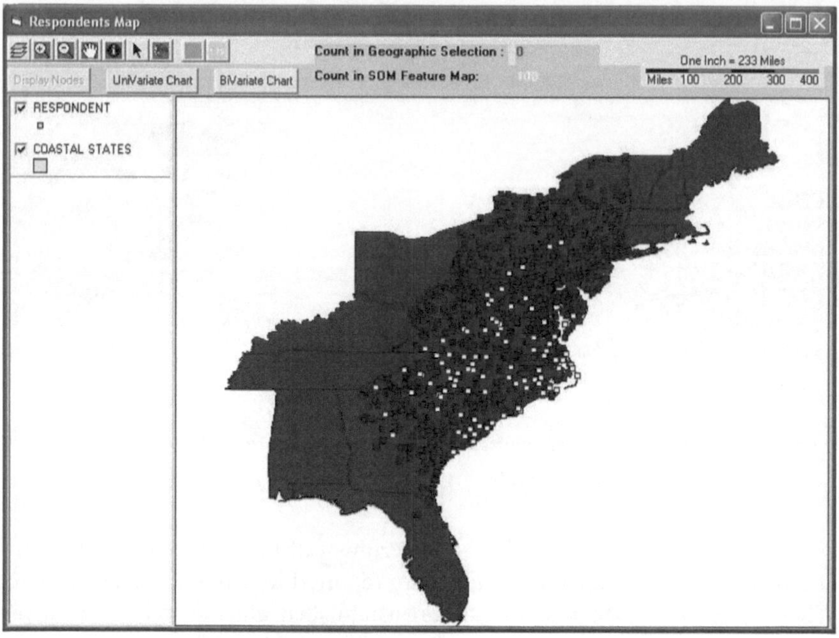

Figure 6.3 The Midland and Southern regions, maybe (*Hearth* set)

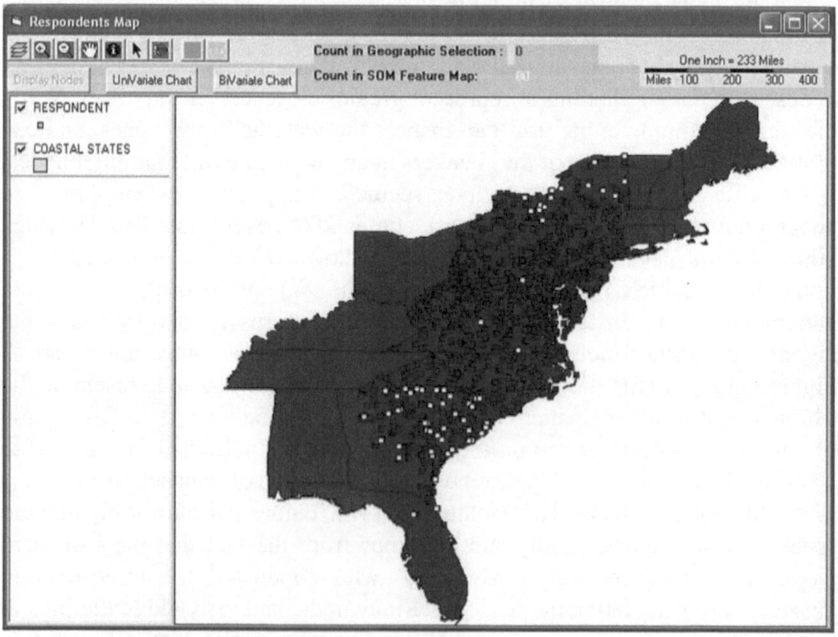

Figure 6.4 Raven McDavid as a fieldworker (*Day/Wind* set)

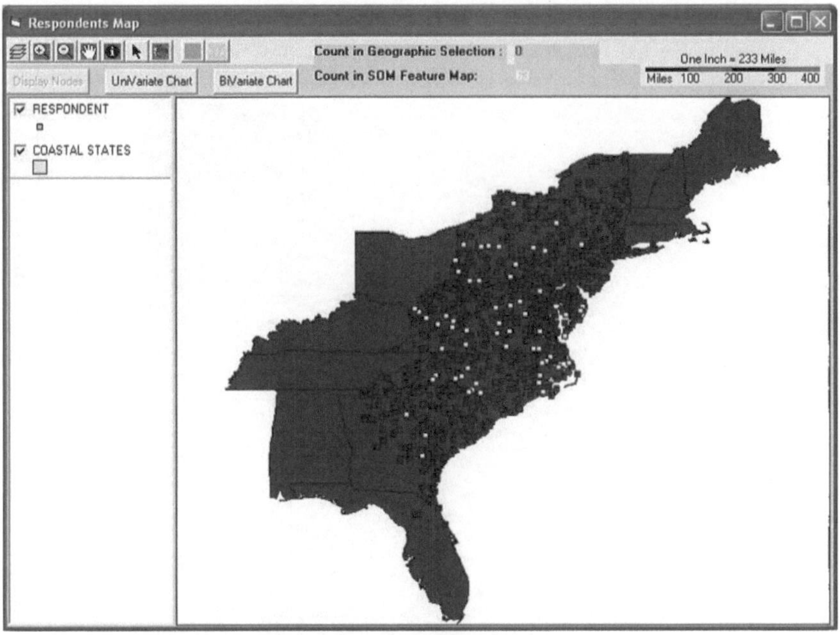

Figure 6.5 Scattered speakers (*Rain* set)

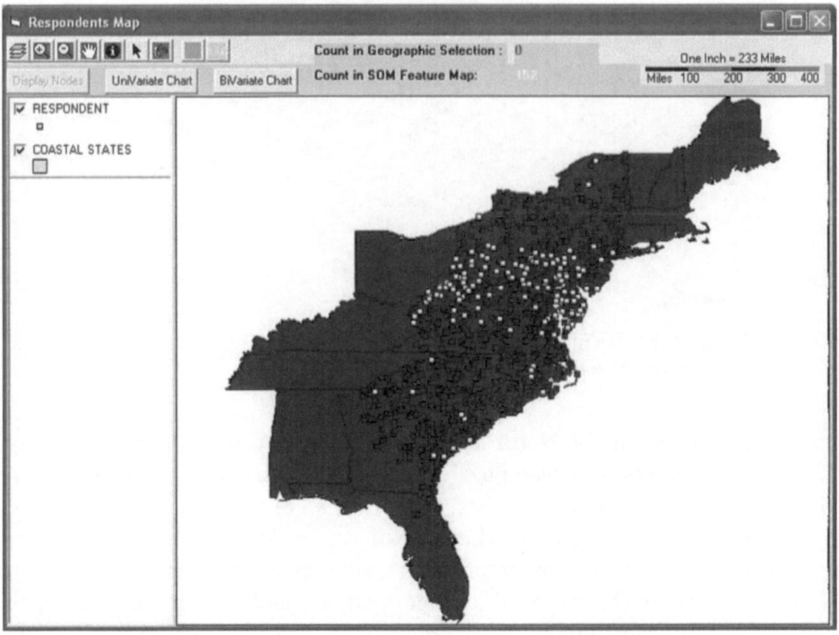

Figure 6.6 North Midland region and outliers, maybe (*Land/Yard* set)

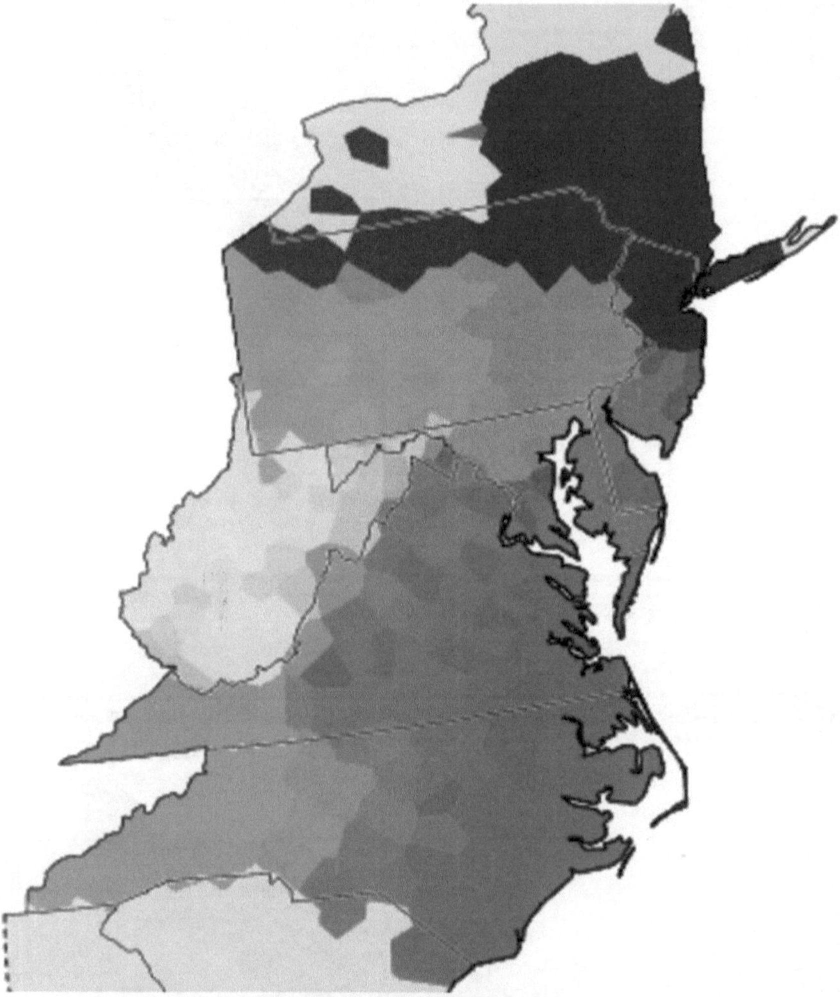

Figure 6.7 Map made with Kleiweg and Nerbonne's cluster analysis, phonetic data (www.let.rug.nl/~kleiweg/lamsas/traditional)

reminiscent of Kurath's North Midland region but with numerous stray highlighted speakers. It is also puzzling to note that the node patterns contain widely varying numbers of speakers, not sets of all of the speakers appropriate to dialect areas. The traditional dialect patterns of American English appear to be better approximated by other means, such as Peter Kleiweg's and John Nerbonne's use of hierarchical cluster analysis and Levenshtein distance (Figure 6.7). Their technique generates areal patterns that can be more easily associated with, while still not exactly matching, traditional regions as drawn

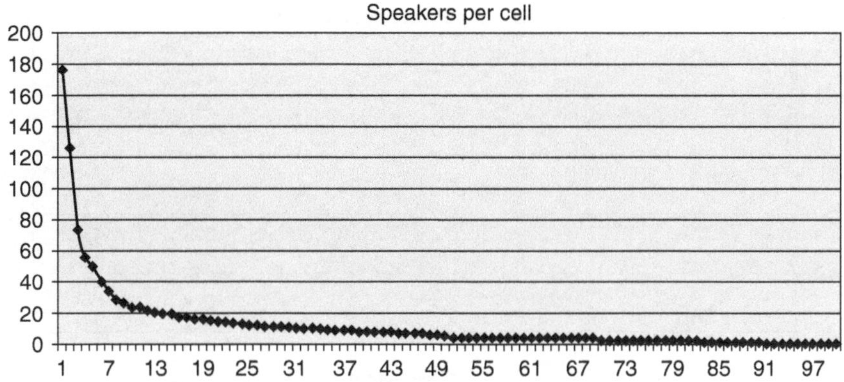

Figure 6.8 Ranked node frequency, 10 × 10 matrix for *Hearth* set

by Kurath and others. Kleiweg and Nerbonne, however, must also worry about confounding factors like field worker differences.

However, we should take a closer look at what our Self-Organizing Maps have actually found by going underneath the colored nodes. First of all, let us look at a ranked frequency chart of the number of speakers in each of the 100 nodes (Figure 6.8). What we see is an A-curve: the neural network algorithm has reproduced in the node values the pattern that we always see in the language data for individual features. We can take this to be a sign that the neural network process really *is* on the right track. Lest anyone think that the *Hearth* set is the only word list with just this distribution, Figure 6.9 shows two more lists of about the same number of words, now for the *House/Inside* and *Land/Yard* sets of words. And for good measure, three more lists, for *Time*, *Day/Wind*, and *Rain* word sets are illustrated in Figure 6.10. They all have an A-curve for the node frequencies. The unguided, massively repeated iterations of the neural network process have led to the sort of emergent order that we expect in language patterns from the operation of complex systems. We might have thought that nodes, and patterns of nodes, would each have a number of speakers that matched some human population, whether a locality in an individual node or a region in a colored pattern of nodes. But the range of sizes for the rest of the nodes is, actually, what we should expect from the scale dependency of the complex system. We have carried out the experiment on one level of scale – the highest level for this survey, all speakers – and we have received one A-curve as a result. The fact that some nodes are empty can be explained by the fact that, somewhere in the iterative process, speakers moved out of them to other nodes. To have some empty nodes is a normal outcome of neural network analysis. The best way to interpret the output of the Self-Organizing Map, the A-curve of speaker frequencies in the nodes, is to say that we have found the top-ranked pattern for this data set, and we have also identified a number of

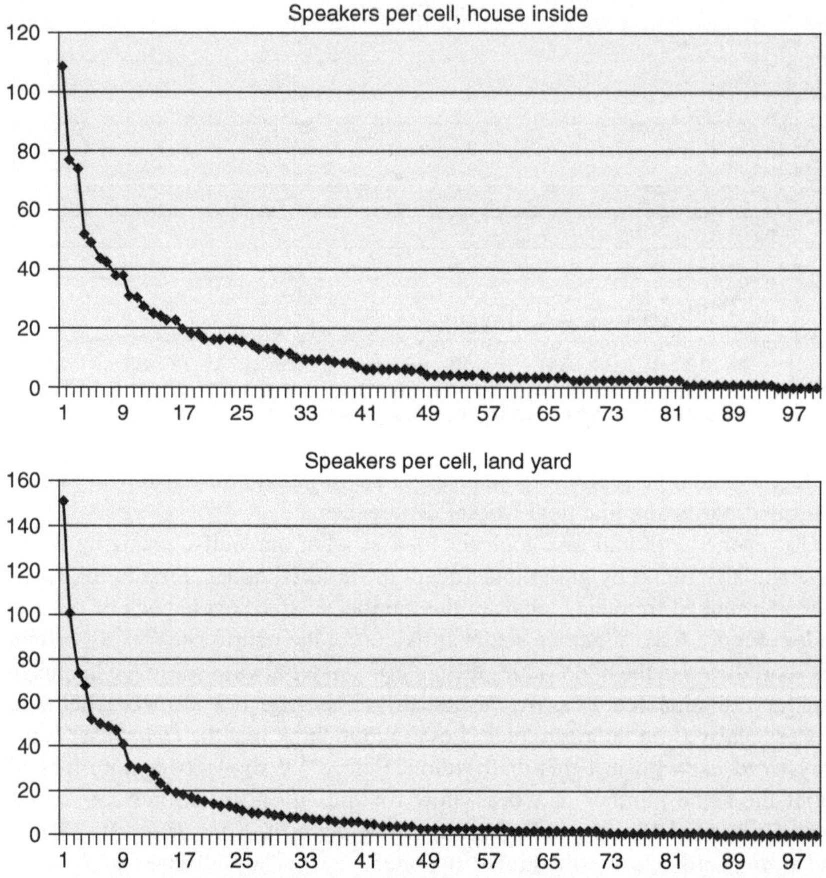

Figure 6.9 Ranked node frequency, 10×10 matrix for *House/Inside*, *Land/Yard* sets

other moderately frequent patterns. We also now know which speakers, for the *Hearth* list at least, do not follow any of the larger patterns. That is, the nodes that have just one or a small number of speakers can be interpreted as outliers to frequent patterns, and most patterns are outliers. Indeed, we can use the 80/20 Rule to characterize our distribution of patterns from the 10×10 array: only 20 percent of the nodes will contain 80 percent of the speakers, and the remaining 20 percent of the speakers will use 80 percent of the possible patterns.

Given our observation of the A-curve in the results, we cannot expect the colored sets of nodes with similar patterns to show us dialect regions. It does not make very much sense to aggregate node frequencies into colored sets because the frequency-ranking yields a single curve, not a distribution that is

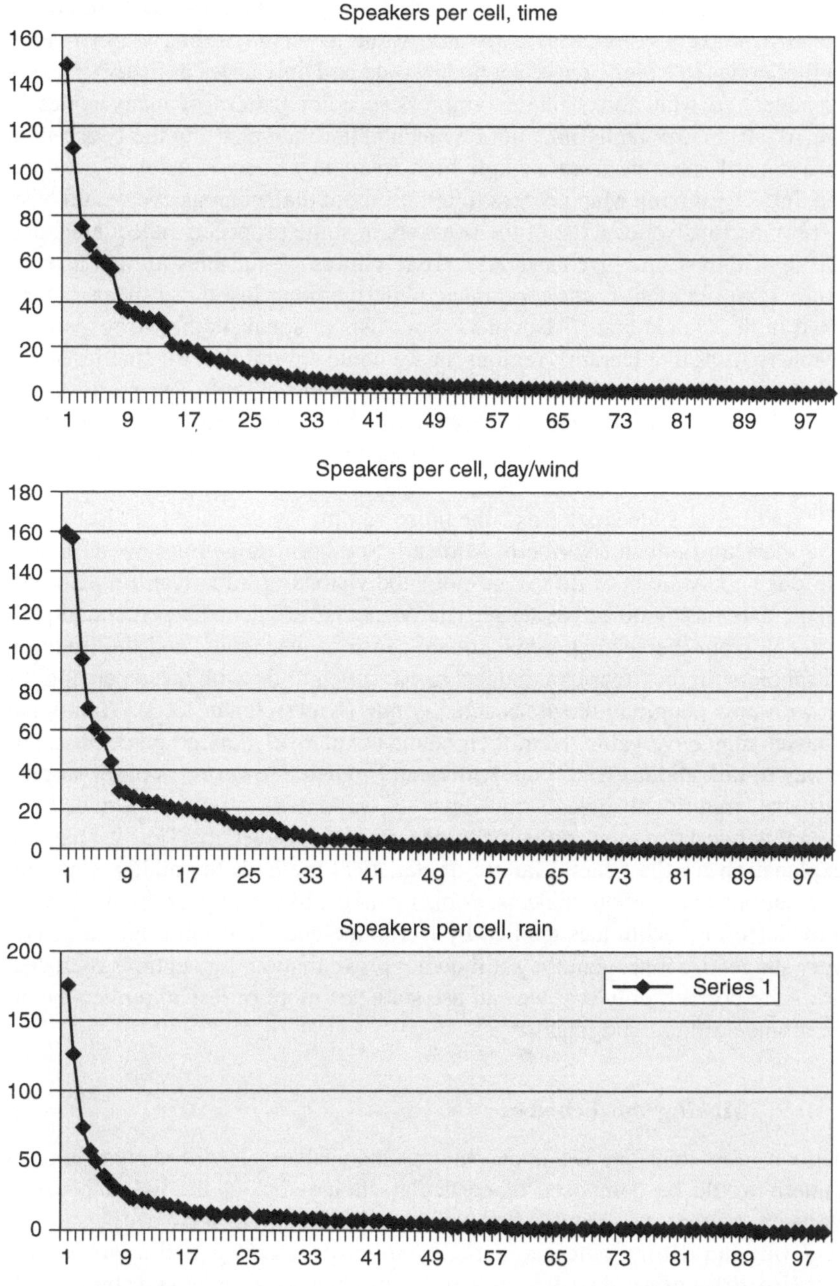

Figure 6.10 Ranked node frequency, 10×10 matrix for *Time*, *Day/Wind*, *Rain* sets

easy to take apart. Associating the low-frequency nodes to the high-frequency nodes to make a colored set adds little value to them. Adding ten or twenty outlier nodes to a high-frequency node would add only about as many speakers as nodes, so what looks like an impressive color pattern of many nodes in the 10×10 matrix really may not be much better than charting the geographic locations of speakers from a single high-frequency node. If we really wanted the Self-Organizing Map process to tell us about dialect areas, we would have to provide in advance a list of the speakers in some proposed dialect area, and the algorithm would give us an A-curve of pattern frequencies for that level of scale. It would identify the top-ranked pattern among just those speakers who lived in the dialect area. We could submit lists of speakers that correspond to whole regions like Kurath's regions, or we could submit lists for smaller areas, say for a metropolitan area. Neural network analysis, however, really is not good for finding traditional dialects blindly across the entire set of speakers because of the nonlinear distribution of node frequencies in the patterns it finds (on attributive and blind dialects, see Kretzschmar 2009: ch. 2).

"Traditional dialect patterns" like those commonly described in discussions of American English (Northern, Midland, Southern, sometimes Western, and various subdivisions of these) are not subdivisions of American English language data in any objective sense. The "dialects" are actually perceptual phenomena from the second, conventional sense of aesthetics, and the tradition of scholarship on American dialects has as much to do with these perceptions as with how people in the areas actually talk (Kretzschmar 2003). Traditional dialects may have value, from their status as relatively shared perceptions, as a way to talk about variation in American English. However, as observational artifacts, traditional dialects can never be validated, only approximated, by neural networks or other statistical processing of speech data. Thus we have an explanation for the general failure of statistical methods to validate what our perceptions tell us about dialects. Some methods, like the Levenshtein distance and clustering techniques applied by Nerbonne and Kleiweg, offer attractive, very suggestive maps, but they still do not prove that our perceptions of dialect areas are correct, only that we can use statistics more or less to replicate conventional aesthetic judgments.

Making word choices

We can now consider the node patterns themselves. Traditionally, a dialect pattern would be composed of particular choices out of the list of possible variants, so that many individual speakers might agree, for example, to say *andiron* (and not the other variants), *mantelpiece*, *backlog*, and *lightwood* for our *Hearth* set of words. Alternatively, we might also consider that some words would not be used in a dialect, so that a choice to use such a word would indicate that a speaker was not from the region. For example, it is probably true that

node	1	10	1	8	3	4	1
	1	1	2	1	7	3	4
andiron							N
dogiron		Y		Y		Y	Y
dogs	Y	Y		Y		Y	Y
firedog				N			
fireiron	Y	Y		Y		Y	Y
handiron	Y	Y		Y		Y	Y
fireboard					Y	Y	Y
mantelboard			Y		Y	Y	Y
mantelpiece		Y			N	Y	N
mantelshelf	Y	Y		Y	Y	Y	Y
mantel			Y	Y	Y	N	
shelf	Y	Y	Y	Y	Y	N	Y
backlog	N	Y		N		N	
backstick			Y		Y	Y	Y
chunk				Y	Y	Y	Y
log		N		Y		Y	
kindling							N
lighterd					Y	Y	Y
lightwood							
pine					Y	Y	
pitchpine					Y	Y	Y
richpine					Y	Y	Y
speakers	176	126	73	56	50	39	33

Figure 6.11 The seven top-ranked nodes

Northerners should not say *y'all* (as in my own case, even having lived for a long time in the South). We can also consider that dialect speakers might just mostly agree, not all agree, but still expect that there should be relatively clear choices of feature variants for us to be able to tell dialects apart. But that is not what the neural network algorithm found. Figure 6.11 shows the *Hearth* variants of the seven top-ranked nodes, arranged by quartiles. There is a "Y" if the weight for the presence of the word was .75 or higher, indicating that the word was selected for the pattern. There is an "N" if the weight for the absence of the word was .75 or higher, indicating that the word should not be used by speakers in the node. I have left the word unmarked if the weight was lower, and thus omitted trends in a node to use a word in order to focus on the most extreme weights. A clear choice, as in our traditional perception, would consist of a Y for one variant, and an N for the others within a single questionnaire cue. In Figure 6.11 we just do not see clear choices in the evidence. The top-ranked node (1, 1) shows two variants selected for the "shelf" cue, and it has three variants selected for the "log holder" cue, only the absence of *backlog* for the "log" cue, and no selection at all for the "fire starter" cue. The third-ranked node (1, 2) may be the closest to the pattern we might have expected since it only has four variants selected out of the twenty-two possible, but three variants are selected for one cue, a single variant for another, and no variant selected for the other two cues. The other five of the top-ranked patterns all show multiple selections of words for cues, and this is the general trend in patterns for the other ninety-three nodes not shown here.

The Self-Organizing Map process, then, is telling us something about the shape of the "best" patterns in the data. There are usually not clear choices between the possible variant words for notions or objects like the *Hearth* list cues. Instead, speakers in the nodes appear to have a repertoire of words for every notion, a pool of choices, although in some cases the absence of a word is selected so that speakers in a node should lack certain words, or should avoid certain words like a Northerner should avoid saying *y'all.* Not saying a word can be salient.[5] Of course, this experiment only used the top-ranked variants for each of the cues in order to run the neural network algorithm – the pool is actually much, much deeper in each case. However, even in this limited experiment which should have made it easier to find clear choices, the top-ranked variants for each notion are selected together for patterns, or are present to some degree in trends as indicated by the empty cells in the table.

This finding suggests that differential frequency is a much more important factor than binary, on/off selection in the way that different words are used differently by speakers. And this is an indirect clue to our cognitive function. That speakers have pools of variants, and do not make clear choices in dialects, is what allows the simultaneous existence of many regional and social patterns in language use, and also many discourse types, all distinguishable on the basis of differential frequency. We all participate in many of these patterns ourselves at different moments in our days and lives. We must have cognitive facilities that

[5] The fact that *not* saying something can be important in dialect recognition, as well as the use of dialect features, raises the cognitive issues of consistency and coherence. Lakoff and Johnson introduce relationships between cognitive metaphors in these terms (1980: 45). Not to say something characteristic of speakers outside one's region would preserve the perception of consistency. However, since it often happens that use of metaphors, and use of dialect features, is not consistent, the notion of coherence can be applied. Inconsistent use of metaphors, or use of dialect features, can be reconciled as coherent if it is possible to see them as entailments of the same superordinate structure. Lakoff and Johnson suggest that "connections between metaphors are more likely to involve coherence than consistency" (1980: 44). The same general point, that clear and consistent choices are less important than differential frequencies that invoke coherence for features, arises below.

support such decision making. Indeed, the connectionist model in neuroscience of massive parallel processing fits this evidence well, while the physical symbol system and neurolinguistics models that rely on prototypes (more similar to on/off selection among alternatives) do not. Parallel processing is much more flexible, in that it allows habituation in many pathways as opposed to a prototypical preference for single pathways. Whether a neural network model or a dynamical systems model fits the data better is a matter of perspective. The neural network model led us to the A-curve of speaker frequency in nodes, and to an array of weights in the choice of words in domain sets. The dynamical systems model suggests, however, that the finding from neural network analysis is not an endpoint to analysis, but instead just one point in a constantly moving complex system of connections and interactions. When we consider the fact that our linguistic experience never stops, our mental model of language around us must continue to change over time in a dynamic process.

The way that neural network analysis builds its patterns provides clues for how we might best put our emphasis when we try to use explicit statistical and scientific means to tell apart and to put to good use the language patterns that normally we just perceive to be out there, often just as a matter of convention. It also suggests how we might use instrumental means more effectively for direct observation of neural function. The experiment has shown us how much an aesthetic difference, between objective perception of speech evidence and what our taste in speech patterns might lead us to see, has controlled our thinking about dialects. What was puzzling before about the way that Self-Organizing Maps failed to show traditional regional dialect patterns has its explanation in complex systems. It is not the implementation of neural network analysis that is the problem, but instead a mismatch of traditional assumptions against the actual distributional patterns in the evidence. The details of the neural network results show us that the algorithm has actually done a good job of reproducing patterns in speech, just the emergent patterns from the complex system of speech and not traditional dialect patterns. We can use neural networks to study traditional dialect patterns, but only if we control the scale of analysis instead of trying to reproduce areas blindly, and only if we pay attention to the nonlinear distribution of the frequencies of the variants included in the analysis. The neural network simulation does give us clues to help understand neural function itself, in that, unexpectedly, it has modeled decision-making in such a way that it suggests how habituation and activation patterns may be involved in linguistic decision-making in the brain. In the end, we still cannot use Self-Organizing Maps to validate traditional dialect areas from the data because these areas derive more from our perceptions as observational artifacts than from clear choices in language production. However, thanks to neural network analysis and a knowledge of complex systems, we can begin to develop important insights about the patterns in our speech data and how we perceive them.

The experiment also tells us something about aesthetics and neuroscience. Archbishop Berkeley long ago taught us to question our perceptions, as when a long pole thrust into a clear pond appears to bend just below the surface. We have seen that the same thing happens with speech: the cause is not a physical property like refraction in the water of a pond, but instead our social construction of the world around us. The fact that we agree about something, like the real existence of dialects, does not make it right in objective terms. Our agreement makes use of the evidence, in this case the fact of language variation, but the agreement itself is a social construction, not necessarily the best or the only characterization of the evidence. As we consider the emerging findings of cognitive science, we should avoid merely asking whether they agree with what we think we already know. It is one thing for an author like Bermudez to offer a balanced perspective about the different cognitive models in the course of a textbook review (his book will be used by cognitive scientists from different camps, and needs to offer a history of the field), and quite another to continue to apply notions that cannot be upheld by the evidence. To continue to support the physical systems model in linguistics answers only to a question of taste, not the evidence that continues to emerge from neural network and dynamical systems experiments. We can of course continue to talk aesthetically about the social acceptability of neuroscience according to our prevailing social construction of the world. This is a question worthy of discussion (nobody has ever really accepted the old saw that *de gustibus non disputandum*), but it should not be the only topic on the table. We should also ask about the philosophical aesthetics of the new neuroscience, how it may help us to see beyond our current taste and create new formulations from objective perception of the evidence.

7 Sociolinguistics, communities, and complex systems

Complexity science tells us about the process by which order emerges from massive numbers of random interactions among the components in the complex system of speech, rather than from a simple cause or set of rules or other controlling agent. Whether we are talking about ants, computers, physics, or speech, a few simple principles can give rise to frequency-based patterns. As we have seen, these patterns are not the same as what we usually call grammar. Instead, complexity science defines the relationship between language in use and any generalizations we may wish to make from it. It allows us, really for the first time, to get away from reliance on our aesthetically conventional perceptions of the language around us, perceptions that are composites of linguistic and non-linguistic information, and instead to make effective use of the actual patterns of language as people use it to resolve practical problems in language study. For the field of sociolinguistics, the area in which collection and analysis of actual speech has played such a strong role, it is all the more important to understand in greater detail the actual patterns of language variation that emerge from the complex system of speech. In this chapter we will examine the foundations of North American sociolinguistics, the idea of the speech community, and the special case of "new varieties" of a language. We will then also consider ways in which the variation observable in sub-populations of different sizes is typically distributed, and how such variation can best be measured.

Classic North American sociolinguistics

North American sociolinguistics is a school, in the academic sense of its practitioners sharing essentially the same assumptions and following the same methods. Bill Labov's brilliant founding insight for sociolinguistics was to account for evident variation in language by proposing the idea of "coexistent systems," parallel grammars that corresponded in the first instance to social "Class" but also to ethnicity, as among the Italian, Jewish, and black residents on the Lower East Side of New York (1966; Weinreich, Labov, and Herzog 1968). Labov also invented the sociolinguistic notion of "Style," the deflection of language performance towards the speech of the highest class, which he identified with Standard English, as observed in different kinds of spoken tasks from minimal pairs at the most formal end of the spectrum down through the most casual speech. Within a few years Labov described how features within such coexisting systems were also subject to form/function variation according to the linguistic environment in which they appeared. Finally, from the beginning Labov conceived of the Class/Style framework and form/function variation as elements of linguistic change in progress, so that the focus of sociolinguistics has always been not so much about the status of the linguistic hierarchy, but more about its continual readjustment in social context. Coexistent systems and variable rules may have violated the elegance of the assumption of homogeneous speakers in the generative model, as Chomsky (1965) might put it, but many linguists have considered this a small price to pay for a chance to account for the obvious evidence of variation that we see all around us.

We should consider on what terms Labov conceived of coexistent systems, which we might also call parallel grammars, and the Class/Style framework. As J. K. Chambers has put it in his book *Sociolinguistic Theory*, "The essential premise of all linguistic research is that language is systematic. If it is not, there is nothing more to be said. Linguistics cannot exist" (2009: 17). Chambers argues that language must be systematic as a sort of contract so that speakers can understand each other, following, in his term, the "axiom of categoricity" (2009: 12–13, 26–37). We see in this axiom the North American bridge between formal linguistics and sociolinguistics. Of Labov, Peter Trudgill has written that (1984: 2–3):

> he actually resisted the term *sociolinguistics* for some time, as this seemed to him to be in danger of opening up the way to a series of correlational studies of little theoretical interest. He would have preferred, that is, to refer to his work simply as *linguistics*. Nor is work of this type particularly concerned with the social conditioning of speech. Rather, it is concerned to learn more about language, and to investigate topics such as the mechanisms of linguistic change; the nature of linguistic variability; and the structure of linguistic systems.

The nature of each parallel grammar, each coexistent system, therefore, is its formal structure, so much so that Labovian sociolinguistics is not "particularly concerned with the social conditioning of speech." For this reason it is a mistake to think that North American sociolinguistics begins with social conditions, though Labov and other sociolinguists have always been interested in social fairness; classic North American sociolinguistics began as, and remains, a formal linguistic discipline.

J. K. Chambers defines the axiom of categoricity as "the simplifying assumption that data for linguistic analysis must be regularized to eliminate real-world variability" (2009: 12).[1] He cites Sapir's 1921 statement from the beginnings of American linguistics that "Everyone knows that language is variable" (2009: 11) but then describes Sapir's practice of treating variation as "loose ends," and cites Sapir's famous aphorism that "no language is tyrannically consistent. All grammars leak" (2009: 11). Chambers claims that "the generation that came after Sapir adopted a very strong form" of this view, and that Chomsky "influentially perpetuated" it thereafter (2009: 12), as in his famous definition of the object of linguistic theory as "an ideal speaker-listener in a completely homogenous speech-community" (2009: 27). Kretzschmar (2009: ch. 2) particularly associates this view with Saussure, who focused on *langue* as 'linguistic structure.' Saussure did recognize perceptual, cognitive, social, and historical information as relevant for the study of language, and yet he thought that trying to consider all of these aspects at once would risk losing control of linguistics as a science. The axiom of categoricity, then, arose historically out of an inability to manage the variables relevant to language and the variability in language that obviously exists in its use, and the axiom has been continued by Chomsky and others as part of a rationalist idealization that considers language variation irrelevant to the work (see Chambers 2009: 27–28).

Saussure's alternative to langue was *parole* 'language in use, speech' (see Kretzschmar 2009: ch. 2, for an extended discussion). Chambers claims that modern sociolinguistics has at least been "relaxing the axiom of categoricity" (2009: 35), and has "made tremendous strides in the development of the science of parole" (2009: 28). While it is certainly true that North American sociolinguistics embraces variation in a way that Chomsky has not, it does not follow that sociolinguists therefore accept *parole* as their object of study. North American sociolinguists have taken an interest in what people say not as personal language behavior, but as members of social classes and ethnic groups, sub-populations that are assumed to exist but have seldom been validated in actual communities. Acceptance of language as an array of coexistent systems allows descriptions of hyper-correction, style-shifting, code-switching, and code-mixing as what speakers may do to negotiate their use of different systems (Chambers and Trudgill 1998: 110–113 also refers to fudged and mixed lects). Whatever the term used, Weinreich, Labov, and Herzog (1968: 165) specify the process that for them accounts for "variability within the system:"

> The heterogeneous character of the linguistic systems discussed so far is the product of combinations, alternations, or mosaics of distinct, jointly available subsystems. Each of these subsystems is conceived as a coherent, integral body of rules of the categorial, Neogrammarian type: the only additional theoretical apparatus needed is a set of rules stating the conditions for alternation.

[1] This section derives from Kretzschmar (2010: 264–266).

Variability, then, consists of "alternations," combinations or substitutions of rules from one system with or for the rules of another. The empirical foundations of sociolinguistics, then, have not relaxed or abandoned the axiom of categoricity, but instead they multiply it so that every subsystem in the array of coexistent systems is bound by the axiom. Classic North American sociolinguistics embraces and correlates empirical linguistic data with social factors, but it does not yet address *parole*; it just maps coexistent systems to perceived social groups. On the other hand, complexity science does address *parole*, on its own terms, as we have seen.

By way of illustration, let us consider the process that Labov used to identify what he thought was genuine New York City speech.[2] The key element of Labov's exposition comes directly from his assumption that different groups of speakers will follow the axiom of categoricity, that they each will have a system. He groups and regroups evidence in an attempt to resolve, by means of new ideas or rules brought forward to motivate the reformulation, problems or paradoxes in the evidence. That is, he redefines the membership of the set of speakers among whom a linguistic change is expected to operate regularly. The procedure can be illustrated from Labov's attempt to defend the use of "observations in apparent time" (1994: ch. 3), a pillar of his later analysis of sound change in progress. He begins an analysis of the distribution of (ay) and (aw) in a sample of all New York City speakers with a plot of a front-mid-back index against the age of the speakers (1994: 51, figure 3.2), and he finds that age accounts for only 5 percent of the variance. He then creates separate plots for New York City black speakers and New York City white speakers (1994: 52, figures 3.3, 3.4), and finds that age accounts for 11 percent of the variance in the white subgroup while the black subgroup "has a different distribution in apparent time from the rest of the community" (1994: 51). He then produces a plot which includes only the New York City white speakers from "the central social classes," which retains seventy speakers and removes fifty-seven speakers from the highest and lowest social classes (1994: figure 3.5); this procedure yields the result that 14–16 percent of the variance comes from age. Labov assessed the procedure (1994: 54):

These studies show that the speech communities in most northern cities[3] are in fact two distinct communities: one white, the other nonwhite. Within the white urban community we typically find a geographically uniform, socially differentiated dialect built upon a homogeneous structural base (Labov [1989]). But not all variables behave in the same

[2] This section derives from Kretzschmar (1996a).

[3] In the previous sentence Labov had cited his own work in Philadelphia to confirm the separation of black and white speech communities, and had cited Bailey and Maynor (1987) as confirming that the separation had been "observed in most other areas of the North and the South" (1994: 53–54). Labov also separates the black and while speech communities in Philadelphia in Labov (2001).

> way for all social groups. Figure 3.5 confirms the general finding of an association between change from below and a curvilinear pattern in the social class hierarchy. Since changes from below proceed first and at a faster rate among the interior social groups, a clear view of the change in apparent time then requires that we examine these innovating groups separately in order to get a clear view of the main direction and rate of progress of the apparent change. At the same time, we lose information on the behavior of the variable among the less advanced social groups. In the progressive refinements of Figures 3.2–3.5 information is lost as well as gained.

This paragraph clearly reflects the notion that progressive redefinition of groups is required for "a clear view" of what is going on in New York City. Once he disposes of the black speakers, Labov assumes that he will find "a geographically uniform, socially differentiated dialect built upon a homogeneous structural base," in line with his model of Class, among the white speakers. Change in apparent time, he claims, will be most visible in evidence from the "interior" social groups, the working-class speakers. He does admit that, in focusing on the interior groups, the analysis loses information about other speakers.

One may wonder why it is reasonable to exclude black speakers, and later the upper- and lower-class white speakers, when by this act the notion of "New York City speech community" that originally guided data collection is fragmented and lost. Even if we grant that black speakers may form a separate speech community, Labov's progressive redefinition has removed nearly half of the white speakers, and so his notion of speech community for even the white speakers is sharply limited. If we apply a complex systems view to this analysis, we can see that Labov has managed the subgroups in his data but has not changed what he claims to be studying. He begins with the top level of scale, New York City. He then subtracts the black speakers so that his results reflect only the white speakers, no longer all of New York City. Finally, he subtracts the upper- and lower-class whites so that he ends up with just the working-class white speakers, an even smaller part of the population of New York City. According to the scaling property of complex systems Labov is entitled to examine whichever groups of New York City speakers he wants, but he is then no longer entitled to say that he is still talking about New York City as a whole. According to complex systems it is simply an error to think that the working-class speakers in New York represent the whole city. Scaling effects for speech have been discussed outside of a complex systems approach by Horvath and Horvath (2001, 2003), who say that we should not expect to apply generalizations at higher levels of scale to lower levels of scale (called the "ecological fallacy" by the Horvaths 2003: 162), and we should not expect any individual fairly to represent the behavior of a locality, or any locality fairly to represent the behavior of a region (called the "individual fallacy" by the Horvaths 2003: 162). Within complex systems, we know that the scaling property produces the same distributional pattern, the A-curve, at all levels of scale, but as the Horvaths have said, we cannot expect to find exactly the same components at exactly the same frequencies at different levels of scale. Labov cannot use results from white working-class speakers to represent any larger level of scale in the city. He has not found a central property of language change by clearing away the distraction of data from other subgroups of New York City speakers.

His defense of the apparent time hypothesis, then, belongs just to the working-class group that he ends up with. Black speakers in New York, and upper- and lower-class white speakers, do not show the same evidence of change in apparent time. According to the complex systems view, change in apparent time is primarily a characteristic of just one subgroup – and in that group accounts for just 14–16 percent of the variance in (ay) and (aw) tokens. Labov's interpretation of the evidence accords with the coexistent systems of groups that he defines according to his Class model, in accordance with the axiom of categoricity. The only way for Labov to reach his conclusion from the evidence is to consider that, if change in apparent time can be shown to some degree for one system, it is a valid linguistic principle for any system (a "language internal" property), including in this case the overall system in New York City. As for his consideration of historical evidence presented in Chapter 5, Labov has tried to illustrate a general principle from a minority of his evidence. However, the findings of complex systems, and also what the Horvaths have shown, will not accommodate such a conclusion. Linguistic structures are observational artifacts, matters of perception subject to aesthetic conditioning, not generative hierarchies, and so they cannot have such language internal properties. Labov's approach cannot begin to address the real problems with grammars caused by the 80/20 Rule, as presented in Chapter 4.

The same kind of problem arises for Labov's notion of Style, connected as it is with Standard English and with "the vernacular." Labov (1984) contains perhaps the clearest statement of his assumptions, or "working principles" (1984: 29–30):

1 There are no single style speakers.
2 Styles can be ranged along a single dimension, measured by the amount of attention paid to speech.
3 The vernacular, in which the minimum attention is paid to speech, provides the most systematic data for linguistic analysis.
4 Any systematic observation of a speaker defines a formal context where more than the minimum attention is paid to speech.
5 Face-to-face interviews are the only means of obtaining the volume and quality of recorded speech that is needed for quantitative analysis.

Labov considers that all speakers change their style, some more and some less, in response to changing conditions for spoken interaction. Beginning in Labov (1966: ch. 4), he created a cline of situations in which speakers paid the most attention to speech (minimal pairs, word lists), moderate attention to speech (reading passages, interview style), and the least attention to speech (speech outside the formal interview, speech with a third person, speech not in direct response to questions, childhood rhymes and customs, and danger of death narratives). For Labov, these situations thus create a linear scale of formality, in which the more formal the situation the more style shifting occurs. The least formal situation, therefore, offers access to the vernacular, the linguistic system least affected by style shifting, and we should therefore expect to find the vernacular in, for instance, danger-of-death stories. Style shifting may occur in relation to Standard English, which is what people expect to use in the most formal circumstances, and which may itself be the vernacular for some speakers. Style is thus connected to Class, in that upper-class speakers are assumed to use Standard English, and speakers in classes below the top class will style shift towards their impression of Standard English, including the possibility for hypercorrection, in situations where they are paying attention to their speech. Finally, since face-to-face interviews are the only way to collect enough data, and since any systematic interview introduces style shifting, we have Labov's famous observer's paradox, that one cannot observe speakers while they speak unobserved. The core of Labov's exposition of Style, then, comes down to the idea that there is one most basic system for anyone's speech, the vernacular, which is the least formal and is opposed to another system, the most formal one, Standard English. As Sali Tagliamonte has explained it (2006: 8):

> The vernacular is positioned maximally distant from the idealized norm [citations of J. Milroy and S. Poplack]. Once the vernacular baseline is established, the multi-dimensional nature of speech behavior can be revealed . . . Thus, the unmonitored speech behavior of the vernacular enables us to tap in to the broader dimensions of the speech community. In other words, the vernacular is the foundation from which every other speech behavior can be understood.

These ideas, as expressed both by Labov and Tagliamonte, derive directly from the axiom of categoricity, the assumption of a most basic system, now accompanied by style shifting in formal settings. Style as a linear aspect of formality combines with Class as a linear arrangement of social groups, so that the classic North American Class/Style chart of variation shows a set of parallel lines as in Figure 7.1. Figure 7.1 shows the percentage of *-r* realization in New York City for the different Class (SEC) groups and for the different Style classifications, from least formal at left to most formal at right. The chart also shows hypercorrection by the middle-class speakers in the more formal Style categories (hyphenated lines), which we can take to represent part of the multidimensional behavior described by Tagliamonte. Style works with Class in a neat synergy, taken to show significant facts about the reflection of social factors in the system of the language.

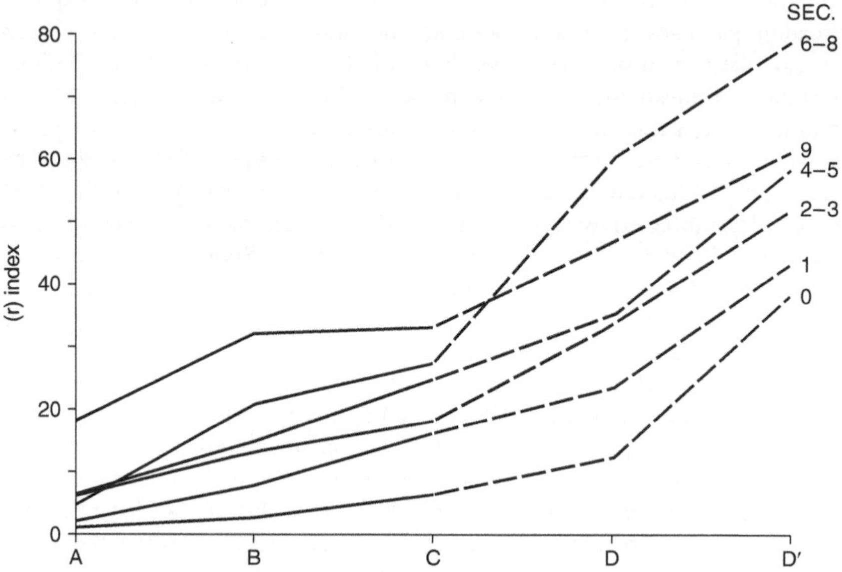

Figure 7.1 Class/Style model for -r in New York City (from Labov 1972b)

For all the neatness of the doubly linear chart, there is good reason to doubt the existence of the vernacular. As Milroy and Gordon point out, "the difficulty in pursuing the vernacular . . . lies with the impossibility of recognizing the quarry when it is caught" (2003: 50). They further suggest that attempting to find the vernacular of any speaker is a "trap" because "it is clear that *any* speech varies considerably in response to situational context," and that "the concept of an entirely natural speech event (or an entirely unnatural one) is untenable" (2003: 50). Moreover, particular attempts to elicit casual speech like the danger-of-death narrative may not work in the same way, or at all, in different populations of speakers (Milroy and Gordon 2003: 65–66). If the vernacular is the foundation for all of the other Style categories, then losing the foundation also loses the rest of the categories. If we cannot recognize the most basic system, we can no longer tell what counts as style shifting.

From the complex systems point of view, each of Labov's Style categories represents a different text type, a different situation for speech, whether a word list or a reading task or an interview. Conversation with a third person or children's rhymes or the danger-of-death story similarly count as their own text types. The interaction of place and group with the textual dimension of speech will have unpredictable consequences, as Milroy and Gordon have said. In short, the Style framework is just too simple to account for the truly multidimensional interactions we all encounter in speech. It reduces the number of possible interactions of groups of speakers and situations of speech to a tiny fraction of what we know to be possible. The idea that both the Class groups and the Style types are both arranged in linear fashion is also an oversimplification, as we will see later in the chapter in the discussion of measurement. As Nikolas Coupland has remarked from his perspective outside North America, "style is a far more elaborate and more theoretically important concept in sociolinguistics" (2011: 140) than Labov's framework suggests. Coupland cites Halliday and Bell, in particular, in illustrating his point, and both of those authors take a much richer approach to style than Labov. Coupland concludes by saying (2011: 156)

> Once we recognize the need for a *performance* perspective on the contextualization of language, we have necessarily given up any reliance on assumptions about "natural" or "authentic" speech. There is no social context where language use proceeds "naturally," free from self-reflexive introspection, strategy, or design, even though different qualities of performative framing will be detected and will influence the meanings that are negotiated.

If language is performed by its speakers according to their own ideas of their situations for speech, rather than just being generated as a consequence of their station in life and the formality of the situation, then the simplicity of the observer's paradox is all the more troubling. Speakers offer much more for us to consider than Labov's framework can address. And, of course, every coexistent system is subject to the same problems for the construction of a grammar as described in Chapter 4, and the axiom of categoricity has not encouraged any such analysis.

What should we make of Labov's achievement from the point of view of complex systems? First of all, we can see that Labov's Class and Style model began to capture both the geographical/social dimension (through Class and ethnic groups) and the textual dimension (through Style) of speech as a complex system. What Labov could not have known in 1966 was the pervasive and unlimited operation of the scaling property of speech in a complex system. His introduction of a framework to handle parallel grammars was certainly a fine beginning, but now we know that no small, fixed set of coexistent systems could account for the multidimensional patterning of language in use from complex systems. The subsequent development of sociolinguistics, to include observation of groups of speakers in social networks and communities of practice, moves in the right direction by recognizing additional groups of speakers, at smaller and smaller levels of scale. Becky Childs, for instance, has written about two communities of practice in North Carolina, each with only four speakers (2005). Complexity theory tells us that there is no conflict between Class and Style, social networks, and communities of practice. The scale-free nature of speech allows social networks and communities of practice to exist concurrently with larger groupings like social class or region, right up to a national variety like British English, American English, or any of the new English varieties that Schneider (2007) has described. Moreover, the notion of coexistent systems, parallel grammars, is a very good match for what we now recognize as the output of complex systems, the frequency-based patterns of variants for every linguistic feature for every community at every scale – so long as we do not artificially limit how many coexistent systems there might be for any language, even for any large community like New York or Philadelphia. Finally, the sociolinguistic emphasis on language change also matches what we now know, that complex systems cannot stop operating and thus changing, or else we end up with frozen, "dead" languages. Still, in making these connections, we should keep in mind that Labov and other North American sociolinguists have not abandoned the axiom of categoricity; they have just displaced it from the top scale of speakers of the language as a whole population to an array of "categorial" systems among sub-populations of speakers at lower steps in scale. So, what Bill Labov started for us has turned out to be essentially in line with what we now know about complex systems. The question now must be how we might study different groups of speakers, and how we might measure their language behavior, with methods more adequate to the multidimensional geographical/social and textual dimensions.

Speech communities and populations of speakers

We can find a richer framework for the geographical/social dimension in the notion of regions, from cultural geography. Wilbur Zelinsky has written extensively about regions as culture areas.[4] He defines a "culture area" as (1992: 112)

> A naively perceived segment of the time–space continuum distinguished from others on the basis of genuine differences in cultural systems. The two characteristics that set it aside from other varieties of geographic region are: (1) the extraordinary number of ways in which it is manifested physically and behaviorally; and (2) the condition of self-awareness on the part of participants.

Thus a region is not merely some arbitrary tract of geography, but instead a location in time and space in which people behave in some particular way, and in which we may find physical evidence related to that behavior. Moreover, the participants have to realize that they are participating in the culture of their area. This is not to say that the culture of an area must necessarily control individual identity. As Zelinsky says, "The recognition of the area and one's affiliation with it can run the gamut from a raging sense of nationhood to

[4] This section derives from Kretzschmar (2011b).

fleeting, unemotional perceptions of small gradations of differences between one's local way of life and another" (1992: 113). But self-awareness, even "if it is present only at the subliminal level" (1992: 113), is required for the definition of a culture area, as opposed to the arbitrary delimitation of some area by scholars or others. Zelinsky points to the use of state boundaries in the continental USA (Alaska and Hawaii do have their own cultures) as examples of the creation of "spurious" or "synthetic" culture areas, having been created in nearly every case by political happenstance rather than with reference to culture. The borders by themselves do not create a "Minnesota" culture, or a "Colorado" culture, even if people do often refer to such a thing; we have to consider the descendants of Scandinavian immigrants and cowboys, respectively, to put the culture into the places. Zelinsky mentions "Chicagoland" as a spurious synthetic region having been created by advertisers. He also cites commercial attempts to promote tourism through creation of simulated historical settings. Zelinsky insists that a region be genuine, and be recognized by its participants, in order for the region to be distinguished from other locations as an area in which we can observe the evidence of culture – of course including language.

If we now shift our attention to language, we can see that region, understood as Zelinsky's combination of cultural elements in a particular area with self-awareness by the participants, defines what first Leonard Bloomfield (in *Language*, 1933) and now more recently sociolinguists would call the "speech community." For smaller groups of people, the sociolinguistic terms "social network" and "community of practice" are not incompatible with this cultural view of region (they can be equated with what Zelinsky has called "voluntary" and "vernacular/perceptual" regions, for groups tied together by cultural circumstances). The potential difference between region and social networks or communities of practice is the inclusion of a spatial variable. In fact, however, location and spatial proximity have been important factors in sociolinguistic studies all along, although they have commonly been left implicit. Labov, for example, has worked in specific locations, first in Martha's Vineyard, then New York's Lower East Side, then neighborhoods in Philadelphia. The Milroys specified different Belfast neighborhoods – Ballymacarret, Clonard, and the Hammer – as the location for the several social networks in their pathfinding studies. Eckert first developed the study of communities of practice in a Detroit-area high school. These particular places certainly affected the results of these studies, even if we commonly focus on their meta-sociolinguistic values and not on the particular people in the particular place. Labov (1963), the Milroys (Milroy 1987), Eckert (2000), and Childs (2005), respectively, do all comment on just this point. Communities of practice may theoretically free themselves from particular locations, an idea developed by Joycelyn Wilson with reference to hip-hop culture in a dissertation (2007), and developed for internet communities in a thesis by Rachel Votta (2007), but most community-of-practice studies treat the language of speaker groups who interact directly, all in the same place. In terms of complex systems, self-awareness of one's membership in a group of speakers in the same place or engaged in the same activity, even if it is subliminal, is what permits the exchange of information, feedback, and emergence of patterns within the group.

Conversely, it is just as important to note that Zelinsky's account of regions addresses a problem in traditional notions of the speech community, which may take just a geographic location as a definition for a speech community instead of indexing location to culture and perception.[5] Just to name a city is not the same as naming a speech community. To say that "Chicago," for instance, is a speech community is not sufficient, and is no better than the advertisers mentioned by Zelinsky who created a spurious "Chicagoland." Richard Bailey (2012) documents the cultural setting of four US cities at specific times in their history, including Chicago, just the sort of approach that leads to description of a genuine Chicago region. Zelinsky's combination of place, culture, and self-awareness provide the criteria with which to determine whether speakers belong to a speech community, and whether they do not belong even though they may live in a place.[6] The geographic boundaries of regions may thus be more inexact than the city limits (like those of the Bible Belt, or of voluntary retirement regions within Florida or Arizona), and still yield a more accurate representation of a speech community than any political boundary. The idea of regions offers a corrective for oversimplified notions of community.

When we consider scale-free networks within the complex system of speech, we are not limited to speech communities with cultural connections and self-aware participants. It is the nature of complex systems to create the same A-curve frequency profiles for variants of features for any population of speakers within the complex system. Thus we could, as noted in Kretzschmar (2009: 56–58), choose an area and describe its frequency profiles for feature variants to produce an attributive dialect. It is unremarkable that dialectologists and

[5] Labov, for example, offers less commentary than he did for Martha's Vineyard on the contribution of place to the language in New York (1994) and Philadelphia (2001). For these cities, as we have seen, he focused on change in progress and the Class/Style model. J. K. Chambers (2003) has argued that such a broad idea of speech community makes it unnecessary to use sampling to collect evidence of language behavior in Toronto. The problem has, however, long been noted by sociolinguists, especially those outside of North America. Suzanne Romaine, for instance in her 1982 book on speech communities, has suggested that there are levels of analysis for individual, network, and social group in addition to taking the speech community as a given, for application of the Class/Style model (1982: 11).

[6] Not everybody in even the most famous culture areas accepts the famous culture. Tamara Lindner (2009) has reported that, in the central parishes of Cajun culture in Louisiana, only sixty to eighty percent of the high-school-age residents she surveyed would call themselves "Cajun" or "Cajun American" – and only 25 percent of residents surveyed in the biggest city in the area, Lafayette, would do so.

sociolinguists could describe special characteristics for any group of speakers they cared to study, because the scaling property of complex systems predicts that there will always be A-curves within the group. Different groups defined by Class will have different frequency profiles for variants, and so will social networks and communities of practice. The fact that special characteristics can be associated with a group of speakers thus does not validate the analyst's choice to study them. The combination of cultural elements and the self-awareness of participants should provide validation for our interest in the language of a speech community, not just its special linguistic characteristics. When we choose speakers for a study, we can just ask them whether they think they belong to a group in which we have an interest. On the other hand, we need not take care to validate a speech community if what we want to do is to construct frequency profiles for feature variants within an arbitrarily defined population of speakers.

New varieties

Edgar Schneider's 2007 book *Postcolonial English* discusses the emergence of new varieties of English in former colonies worldwide.[7] His "Dynamic Model" suggests five phases in the evolution of such varieties: foundation of the colony, stabilization around the outside norm, nativization, formation of an internal norm, and diversification. His description of the histories of English in a number of places, including the USA, shows that the emergence of these postcolonial varieties does seem to follow a similar trajectory. His Dynamic Model fits within the model for complex systems, and an understanding of speech as a complex system can resolve problems in the emergence of what we perceive to be linguistic systems, whether in postcolonial or more traditional populations of speakers.

Here is the central issue, as Schneider has cast it: "the forces of accommodation are effective in both directions and in both [settler and indigenous] communities, and result in dialect convergence and increasingly large shared sets of linguistic features and conventions. Ideally, the end result is the emergence of a single, overarching language community with a set of shared norms. In practice . . . such a high degree of uniformity is usually not reached . . . smaller, socially and ethnically defined speech communities coexist and allow for internal variability under a common roof" (2007: 32). Schneider's Dynamic Model is focused on the one new national variety that he expects "ideally" to emerge in the new postcolonial state. And yet Schneider allows for the presence of coexistent systems. Of the initial formation of internal norms in phase 4, he says "It is characteristic of this phase that the new indigenous language variety

[7] This section derives from Kretzschmar (in press).

is perceived as remarkably homogeneous . . . but not infrequently it is strongly an identity-driven discourse construct. Whatever linguistic heterogeneity remains . . . will tend to be downplayed or ignored" (2007: 51). In the final phase which is supposed to be characterized by differentiation, Schneider says "This differentiation has to be compared to the putative homogeneity of the previous stage, and, given that this homogeneity was found to be valid not in an absolute sense, the same applies here" (2007: 54). So, in order to explain the coexistence of multiple varieties of a language and the emergence of new subvarieties, we require a process for variety formation that does not inexorably lead to a single new variety but instead explicitly allows for linguistic systems to exist in the same place at different scales of analysis. Schneider's "forces of accommodation" cannot do this because, as he says, they would lead to a homogeneous result that does not occur in practice, unless through politically motivated perception leading to "an identity-driven discourse construct." Thus we also need a process that explicitly recognizes the role of perception of varieties, whether to downplay variation or to recognize that different regional and social groups do in fact have different linguistic systems. The complex systems model does just these things.

In the first phase of the Dynamic Model the settler population becomes established in a place that already had an indigenous population, and both groups contribute linguistic features to what Schneider calls, following Mufwene (2001), a feature pool. Schneider and Mufwene say that particular features are eventually "selected" as parts of the system of the new variety. The complex systems model, however, tells us that the feature pool is always with us in speech, and that particular features may be top ranked on the A-curve or may occur at much lower frequencies in the long tail of the curve. While Mufwene's application of ideas from evolutionary biology as a metaphor for linguistic change is not incompatible with complex systems, which has been applied fruitfully in ecology and evolution as reported in Chapter 5, we are entitled to ask what evolutionary "selection" might mean in the context of the A-curve distributional pattern. As discussed in earlier chapters, there is no particular linguistic reason to privilege the most common variants as having been "selected" and therefore to have status as being "systematic" or "phonemic," and to relegate less-common variants just to the status of noise in the system. All of the variants on the A-curve are actually just as relevant for inclusion in the system. The notion of language-internal selection, then, cannot operate within the complex systems model because nothing is ever really chosen or preferred; even loss is rare. What is truly stable and systematic about speech is the curve itself, not any perceived system of selection of variants.

What does emerge from the feature pool within a community of speakers under the complex system model is the set of top-ranked variants on the A-curves for every imaginable element of the language, sounds, words, grammar. It is not the case that the most frequent input variants from the settlers' language automatically become the top-ranked variants in the colonial language. Random interactions between speakers may eventually promote variants at lower frequencies to the top rank, and vice versa, so there is no fixed relationship between input frequencies from settler, indigenous, or adstrate languages and what will become most common in a new variety. The language itself is not all that is involved in the interactions: human agents deploy features in part according to their perception of the relationship between possible linguistic features and the conditions of their use. Complexity scientists use the notion of "positive feedback" to create and manage computer simulations of complex systems (Miller and Page 2007: 51). Moreover, LePage and Tabouret-Keller anticipated use of the word "feedback" in their seminal 1985 work *Acts of Identity* to help define what they called "focussing" for social groups (181–182). Positive feedback for speech consists of speakers' reaction to the relative success of the use of some feature in an interaction, which encourages the speakers to use the feature again. Positive feedback creates the A-curve for every feature, and speakers recognize and respond to the top-ranked feature variants. If we consider communities of speakers, we can see that we as speakers make a connection between our perception of the relative success of possible linguistic features under different conditions of their use. We as agents using language pay attention to our perceptions of what features tend to be top ranked in particular situations, or in other words, we pay attention to what people usually say in local areas, or in social groups or situations, or in communities of practice. The locus for varieties, then, the population of the speech community in which we can expect to find some characteristic variety of speech, is not just the entire colony but all of the situations in which speakers use language. Just as positive feedback creates the A-curve for every feature in a speech community overall, it also creates the A-curve for every feature in every local and every social situation, and we recognize and respond to the top-ranked feature variants in every situation. The operation of the complex system, then, both massive numbers of random interactions between speakers and also the perceptions of the human agents using language, offers a sufficient explanation for how the feature pool actually works, and how it keeps on working over time.

Let us return, then, to Schneider's Dynamic Model. The special value of Schneider's book is that, even though his analysis is not based on complex systems, he consistently observes the limitations that the complex systems model imposes on a formal analysis. He explicitly locates his postcolonial systems at the national level of scale, and so avoids the problem of an undefined population. Schneider is exactly right when he says that "Ideally, the end result is the emergence of a single, overarching language community with a set of shared norms." What Schneider calls "a high degree of uniformity" or "homogeneity" can only ever exist in the idealized abstraction of a linguistic system, not in the complex system of speech. Schneider consistently admits that variation continues to exist at different levels of scale in the earlier phases, and highlights differentiation of lower-scale varieties in the last phase. Best of all, the heart of Schneider's treatment of postcolonial varieties is his study of perception, not just interesting new words and constructions. He considers that the new variety is "an identity-driven discourse construct" in the new speakers' perception, which should suggest to us that linguists' perceptions also create such constructs. In Schneider's initial description of the Dynamic Model, the discussion of each phase follows the same set of sections: sociopolitical background, identity constructions, sociolinguistic conditions, and only then, finally, linguistic effects. His reader can understand the developmental phases, then, not as landmarks in the internal linguistic history of a new variety, but as the evolution of the new society's perceptions. That is as it should be, to accord with the nature of speech as a complex system.

Understood in this way, new varieties are "emerging" all around us every day, as speakers form new groups in local neighborhoods, communities of practice, and social settings, not just in places around the globe with new political or commercial importance. It is not a process that happens once and is done. The complex system of speech continues to operate, and new order emerges from it all the time. What actually makes "new varieties" of English or of any language, in the sense we usually mean in linguistics, is that linguists from time to time choose to record their perceptions of the usage of some population of speakers. We describe linguistic systems not because they are objects waiting for us to find them, but because we grant importance to a particular population of speakers at a particular moment in time. That is what Schneider has done for postcolonial varieties, and with good reason because these postcolonial situations share phases of perception and development. Still, these and other new varieties that we choose to describe may be new to the linguistic literature, perhaps, but they are really just moments extracted from the long-term process of interactions in the complex system of speech. We can find new varieties wherever we can identify a population of speakers, whether we validate the population with reference to new political realities or just by asking the speakers if they belong to it.

An experiment on scale-free communities

Whether we study validated speech communities or arbitrarily defined populations of speakers, we should be interested to understand how speech production in different communities of speakers actually differs. We should say "actually differs" because all of us perceive that different speech communities and populations of speakers use different words or pronunciations. We think that American Southerners say *y'all* and other Americans do not, or that

Canadians say [həus] and [zɛd] while Americans say [haus] and [zi]. These common perceptions are not confirmed by the evidence of speech production, which is never so categorical as we may think, or as the axiom of categoricity tells us. Southerners use *y'all*, and they also use other second-person plural pronouns like *you guys* and plain *you*. Some Canadians say [zi] in what J. K. Chambers has called the "Sesame Street effect," and some Americans say [həus]. The scaling and distributional properties of speech account for how communities differ in speech production. They suggest how our perceptions of differences arise from our experience with speech, and account for how certain differences can rise to "salience."

In an experiment using LAMSAS data,[8] we analyzed lexical data from eight different subsets of speakers for each of twelve different LAMSAS lexical items, selected from different cultural domains. These domains are appropriate to women and men from any socioeconomic group; as Hans Kurath observed, "Regional and local expressions are most common in the vocabulary of the intimate everyday life of the home and the farm – not only among the simple folk and the middle class but also among the cultured" (1949: 9–10). Of course Kurath was describing US culture of the 1930s and 1940s, so we would not choose the same items today. The data files were downloaded from the LAMSAS website and imported into spreadsheet format; any responses that seemed out of place or anomalous, and any "NA" and "NR" responses (missing data codes in LAMSAS), were removed. We then sorted, tallied, and charted the responses, as for *hog pen* in Figure 7.2. The number of different variants for each of the twelve items ranged from a low of 39 variants for *dry spell*, to a high of 239 variants for *cobbler*. The spreadsheets were then resorted for tallies according to eight sub-groups of speakers, by sex, in thirds by age, and roughly in thirds by states, common social and regional criteria by which speech is often grouped and analyzed. Of course other divisions into sub-populations are possible, such as by metropolitan area or education or occupation, but sex, age, and state are familiar categories and adequate to demonstrate scaling. We have not, of course, defined speech communities in these divisions, because we have not tried to associate cultural elements with the populations of speakers or tried to determine whether speakers were aware that they were members of these groups. The division by age yielded groups of ages 16–60 (an old "young" group), 60–75, and 75–100, and the division by states created a Northern Tier, a Midland Tier, and a Southern tier, not quite equivalent to the traditional division into Northern, Midland, and Southern regions. Figure 7.3 shows the tally and chart for the *hog pen* data for Age Group 1, the younger speakers. The charts for Figures 7.2 and 7.3 show an A-curve in each case. We have reported only

[8] The tabulation for the experiment was conducted by Samantha Knoll and first reported in her undergraduate honors thesis, Knoll (2009).

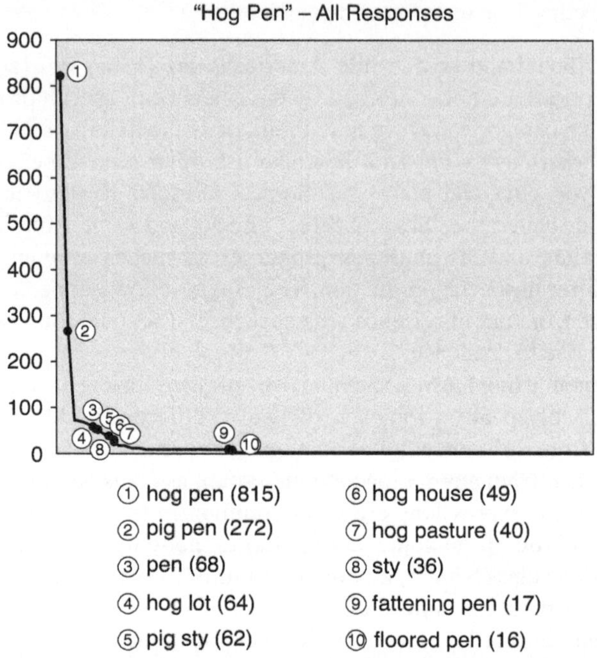

Figure 7.2 *Hog pen*, top ten responses and chart of all responses

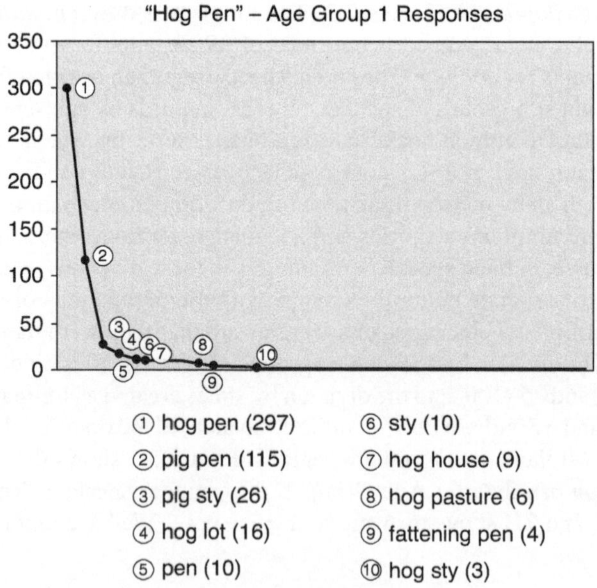

Figure 7.3 *Hog pen*, only Age Group 1

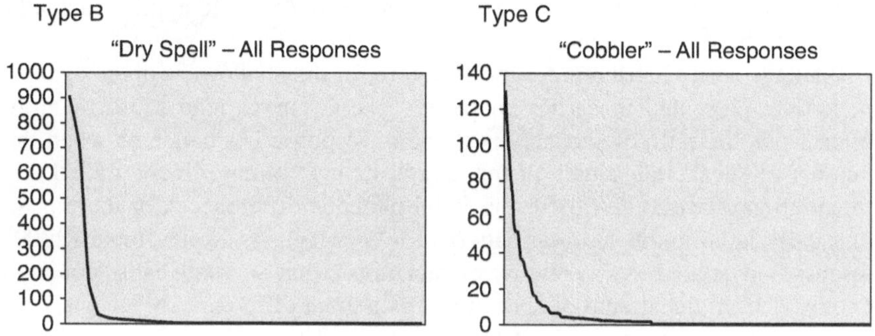

Figure 7.4 A-curves of Distribution Type B and Type C

the top ten response types in each figure, not the large number of infrequently occurring types in the long tail of the curve.

It is evident even from just a cursory examination of the frequency profiles for these twelve lexical items overall and in their sub-populations that a definite pattern emerges. While the graphs vary in minor ways, every single chart of the 108 we made (one chart overall and for each of eight subsets, for twelve LAMSAS cues) has the form of an A-curve. Within this overall consistency, we found that there were three different types of A-curve, called Distribution Types A, B, and C. Curves with Distribution Type A are characterized by the presence of a single highly frequent variant, several variants with moderate frequencies, and many variants with very low frequencies. This distribution results in the A-curve with a long tail that we have seen repeatedly in earlier chapters. Type A curves have a huge drop between the top-ranked variant and the next in order, such that the top-ranked variant is three to six times more common than the second-place variant, and not infrequently up to ten times more common. Of the 108 charts, 84 (78 percent) have the Type A distribution. Distribution Type B occurs when the first two variants in the data set have approximately equal frequencies. The curve for this type begins with a "bump" or gradual slope at the top, then drops rapidly and ends with the long tail. Type B accounted for 15/108 (14 percent) of the charts. Distribution Type C is characterized by the presence of three or more responses with high frequencies, followed by a drop in frequency and the presence of more sets or pairs of variants with similar frequencies, a pattern that continues into the long tail of the curve. The *cobbler* item, which had the largest number of different variants in the experiment (239), is the only one of the twelve items to exhibit this type of distribution (9/108, 8 percent). Still, as evident in Figure 7.4, Type B and Type C distributions do not change the essential asymptotic hyperbolic shape of the curve. Finally, when Distribution Type A is considered in comparison to Distribution Types B and C, we see that the 80/20 Rule applies (actually 78/22), perhaps not an unexpected outcome in the complex system.

We have seen Distribution Type C before, in the shallower curves of phonetic data presented in earlier chapters. Type C curves tend to occur more often when the ratio of speakers per type of response is smaller, as when the number of speakers is small, or the data is subject to more finely graded differentiation of variation (as for the small phonetic differences), or there is an unusually large number of variants (as for *cobbler*). As Kretzschmar (2009) discusses in detail, the way that we extract units (what Saussure calls "concrete entities") from the stream of speech makes a great difference in our analysis. It has been the general habit of linguists to limit the number of variants we consider for any linguistic feature: we commonly talk about binaries ("grammaticality judgments"), or small numbers of possible realizations for particular features, whether grammatical, phonological, or lexical. A famous example is Bill Labov's department store survey of whether, in a rapid elicitation, speakers used *-r* or not. Such a practice makes it impossible to observe the A-curve distribution, which depends upon the relative frequency of large numbers of possible variants. Another way to test for *-r* in New York City, for instance, would be to consider a large number of words with *-r*, or to plot transcriptions or acoustic phonetic information that shows a wide variety of actual realizations of the *-r* segment (as in Kretzschmar and Johnson 1993). Milroy and Gordon usefully discuss management of the range of possible variation (2003: 152–154). Great variation in possible realizations of features is an assumption borne out in the evidence for lexical and phonetic data in this volume, and explored further later in this chapter.

We can also look at the distributional facts in another way, the order of the variants on the A-curve. In Figure 7.5 we see the top ten variants for the *meadow* item across the subsamples. This is a typical chart, evident in six of the twelve items selected for the experiment: the overall ranking at left and all of the subsample rankings follow the Type A distribution, and a single variant, here *meadow*, sits at the top of all nine lists. In the second rank, occurring much less frequently, there are three variants across the subsamples, mostly *swale* but also *meadowland* and *savanna.* Further down in the list we see a large number of different variants, most of them shared between different lists but in a different order in each list. Thus, we see that it is common, if not typical, for there to be a single variant that is top ranked in all the A-curves, massively more common than any other variant for the same item. However, at the same time, it is typically the case that there are many other variants, mostly the same ones in different subsamples, and they have different relative frequencies and thus different orders on the A-curve. How these populations of speakers actually differ, then, is in the relative frequency of use of variants. Our perception of categorical differences between populations of speakers, of common variants unique to particular populations, is simply not supported by

All	Youngest 1/3	Middle 1/3	Oldest 1/3	Men	Women	Northern Tier	Middle Tier	Southern Tier	Variants
meadow	meadow	meadow	meadow	meadow	meadow	meadow	meadow	meadow	1
swale	swale	swale	swale	swale	swale	swale	meadow land	savanna	3
savanna	savanna	savanna	savanna	savanna	hayfield	hayfield	bottom	bog	4
hayfield	hayfield	meadow land	hayfield	hayfield	savanna	swale grass	bottom land	prairie land	6
meadow land	bog	hayfield	prairie	prairie	meadow land	beaver meadow	pasture	pasture	6
prairie	pasture	prairie	bog	meadow land	pasture	prairie	branch	meadow land	5
bog	bottom	pasture	meadow land	bog	prairie	field	buffalo wallow	savanna land	8
pasture	prairie	field	savanna land	pasture	bog	hay lot	flats	old field	8
savanna land	lowland	bog	swale grass	savanna land	canebrake	meadow land	glade	canebrake	7
old field	canebrake	flat	pasture	old field	lowland	pond	grass bottoms	pasture land	8

Figure 7.5 Distribution Type A order of variants: *meadow*

All Type B	Youngest 1/3 Type B	Middle 1/3 Type B	Oldest 1/3	Men	Women	Nortern Tier	Middle Tier	Southern Tier	Variants
Cloudburst	Downpour	downpour	cloudburst	cloudburst	downpour	cloudburst	downpour	downpour	2
Downpour	cloudburst	cloudburst	heavy rain	downpour	cloudburst	downpour	cloudburst	cloudburst	3
heavy rain	heavy rain	heavy rain	downpour	heavy rain	hard rain	heavy rain	hard rain	gully washer	4
hard rain	hard rain	hard rain	hard rain	hard rain	heavy rain	heavy shower	heavy rain	heavy rain	3
gully washer	gully washer	gully washer	gully washer	gully washer	gully washer	hard rain	gully washer	hard rain	2

Figure 7.6 Distribution Type B order of variants: *cloudburst*

the speech production evidence. People mostly say the same things in different places; they just say them at different frequencies of use.

A-curves of Distribution Type B have two high-frequency features. Figure 7.6 shows the top five variants for the *cloudburst* item (henceforth only the top five variants are displayed, since the next five always show high variability). The overall list and the lists for Age Group 1 and Age Group 2 each qualify as

All	Youngest 1/3	Middle 1/3 Type B	Oldest 1/3 Type B	Men	Women	Nortern Tier	Middle Tier Type B	Southern Tier Type B	Variants
andirons	andirons	andirons	andirons	andirons	andirons	andirons	andirons	firedogs	2
firedogs	dog irons	firedogs	firedogs	firedogs	firedogs	handirons	dog irons	andirons	4
dog irons	firedogs	dog irons	dog irons	dog irons	dog irons	dog irons	firedogs	dog irons	2
handirons	dogs	handirons	handirons	handirons	handirons	firedogs	handirons	dogs	3
dogs	fire irons	dogs	fire irons	dogs	dogs	dogs	fire irons	fire irons	2

Figure 7.7 Distribution Type B order of variants: *andirons*

Distribution Type B, while the other subsamples are Type A. *Downpour* and *cloudburst* alternate between first and second position among the lists, half each way. Both variants are in fairly common use among all of the sub-populations whether or not they are close to equivalent in relative frequency as they are in the Type B curves. The situation is a bit more complicated with the *andirons* item (Figure 7.7), where the lists for Age Group 2 and Age Group 3, and for the Middle and Southern tiers of states, each qualify as Distribution Type B. *Andirons* and *fire dogs* are often in the first and second ranks, but *dog irons* also rises to high frequency, as does *handirons* in one case. In these Type B distributions, there is no single typical "normal" variant for all of the lists. The one item with Distribution Type C, *cobbler*, goes even further in this direction (Figure 7.8). The overall list and all of the subsample lists have several high frequency variants, and there are several different variants at the top rank in the subsamples. While *apple cobbler* is the top-ranked overall variant, *apple pie*, *family pie*, and *apple dumpling* are in the top rank in different sub-populations. If we stand back and consider the evidence from all of these figures, we can see that the order of variants in the different Distribution Types describes the continuum of possible realizations of the A-curve: the movement along the continuum from the sharper, deeper curves of Type A to the shallower curves of Type C has its parallel in greater and greater variability among highly ranked variants across subsamples, not just variability in the lower frequency variants.

So, how do these findings address the relationship between actual differences between populations of speakers and our perception of differences? The effect on our perceptions of the ever-present A-curve is that the top-ranked variant receives a strong perceptual advantage. Every distribution at each level of scale

All Type C	Youngest 1/3 Type C	Middle 1/3 Type C	Oldest 1/3 Type C	Men Type C	Women Type C	Northern Tier Type C	Middle Tier Type C	Southern Tier Type C	Variants Type C
apple cobbler	apple cobbler	apple pie	apple pie	apple cobbler	family pie	apple dumpling	apple cobbler	apple pie	4
family pie	family pie	cobbler	apple dumpling	cobbler	apple cobbler	apple pot pie	cobbler	family pie	5
cobbler	cobbler	apple cobbler	apple cobbler	apple pie	cobbler	deep dish apple pie	family pie	peach pie	6
apple pie	apple pot pie	family pie	family pie	apple dumpling	deep dish apple pie	apple grunt	cobbler pie	cobbler	7
apple dumpling	deep dish apple pie	apple dumpling	pot pie	family pie	apple pie	birds nest	apple pot pie	apple cobbler	8

Figure 7.8 Distribution Type C order of variants: *cobbler*

has an A-curve of variants, which means that there is a top-ranked variant overall, but also top-ranked variants, not necessarily the same ones, in every subsample. The perceptual advantage thus helps us think that, for any linguistic feature, there is a variant characteristic of every sub-population of speakers, no matter how large or small the sub-population. Moreover, Type A distributions confer the strongest perceptual advantage, because Type A top-ranked variants occur three, four, five, even ten times more often than the closest competing variant. Since we know that Type A distributions occur most frequently, about 80 percent of the time, the combination of the distributional pattern with the perceptual advantage can lead us to expect, not just that speech communities and populations of speakers will differ in their top-ranked variants, but that "normal" variants will be present all the time, for every feature in every population. The prevalence of A-curves with Distribution Type A explains why we tend to perceive speech categorically.

However, subsamples always share variants, and even when items have the same top-ranked variant across subsamples, they can always be differentiated by the relative frequencies of the shared variants further down the A-curve. Thus, even though we now know why members of the public might perceive speech in categorical terms, as linguists we are misled if we allow ourselves to do the same. About 20 percent of our items possessed Distribution Type B and Type C, and these lack clearly differentiated top-ranked variants, so we linguists are even more misled if we extend our categorical thinking to those Distribution Types. The A-curve distributional pattern always gives us a top-ranked variant for every feature in every sub-population, and yet we still do need to consider the entire curve, not just its peak. We are quite right to think that every speech

All	Youngest 1/3	Middle 1/3	Oldest 1/3	Men	Women	Northern Tier	Middle Tier	Southern Tier	Variants
pallet	pallet	Pallet	Pallet	pallet	pallet	bunk	Pallet	Pallet	2
bunk	Bunk	Bunk	bunk	bunk	bunk	featherbed	Bunk	Lodging	3
lodging	Lodging	Lodging	mattress	mattress	lodging	shakedown	Lodge	mattress	4
mattress	Mattress	trundle bed	featherbed	lodging	trundle bed	straw tick	Spread	trundle bed	6
featherbed	bed on the floor	Featherbed	trundle bed	featherbed	mattress	bollick	bed on the floor	featherbed	5

Figure 7.9 Exceptional Distribution Type A variant: *bunk* for *pallet*

community and population of speakers will be different, at any level of scale; we just need to consider the whole range of variation to prove it. Some might argue that the prevalence of Type A distributions is a good reason for linguists to prefer analyses featuring categorical variants. However, we should remember that even in Type A distributions the 80/20 Rule, as applied to the frequencies of the different variants, tells us that a large number of variants for any feature still help to account for the majority of tokens, not just the single most frequent variant.

What variants, then, do we tend to notice? What of salience? The best possibility will be variants that stand out from the usual patterns offered by the A-curve. Figure 7.9 shows one possibility in the *pallet* lists. The Northern Tier has *bunk* for a top-ranked variant, and it has a Type A distribution (104 tokens vs. 9 tokens for the next highest variant). People may well notice it in comparison to other groups, because *pallet* is top ranked in other subsamples and in the overall list. Another possibility occurs in the *corn bread* lists (Figure 7.10). *Corn bread* is the top-ranked variant overall and in most of the subsamples, but in the Northern Tier *johnny cake* occurs in the top rank, and *corn bread* is in second place a few tokens fewer. The Northern list is actually Distribution Type B, but *johnny cake* still breaks the expectation that items typically have just one "normal" variant across subsamples. To what degree *bunk* and *johnny cake* were actually salient for speakers at the time of the LAMSAS survey we cannot say. Hans Kurath did not mention *bunk.* He said that *pallet* was characteristic of the Midland and South, and *pallet* is the top-ranked feature in those tiers, but *pallet* is also the top-ranked feature overall. Kurath did select *johnny cake* as one of eight features "characteristic" of the Northern region

All	Youngest 1/3	Middle 1/3	Oldest 1/3	Men	Women	Northern Tier Type B	Middle Tier	Southern Tier	Variants
corn bread	corn bread	corn bread	corn bread	corn bread	corn bread	johnny cake	corn bread	corn bread	2
corn pone	corn pone	corn pone	corn pone	corn pone	corn pone	corn bread	corn pone	corn pone	2
johnny cake	johnny cake	johnny cake	johnny cake	johnny cake	johnny cake	corn pone	pone of corn bread	pone	4
pone	pone	pone	pone	pone	pone of bread	pone	pone	pone bread	3
pone of corn bread	pone of corn bread	pone bread	pone bread	pone of corn bread	pone	corn cake	pone of bread	bread	6

Figure 7.10 Exceptional Distribution Type B variant (close to Type A)

(1949: 12, 14), as against *corn bread* and *corn pone.* He referred to *corn bread* as the "national term" (1949: 67), and we find it in the top rank overall. *Corn pone*, he wrote, was "widely used" in western Pennsylvania and the Atlantic states to the south, while *corn bread* was "gaining ground" (1949: 67). The A-curve rankings do not quite bear out this regional judgment. Salience may well be linked to a large difference in the relative frequency of a variant for a single sub-population in comparison to the order of variants on the A-curve in other communities, especially if the salient variant has achieved the top rank. But not necessarily. Since salience is inherently perceptual, Kurath may have been influenced by his historical knowledge or by other factors in his perceptions about these feature distributions. More recently, Johnstone and Kiesling (2008) have pointed out that local people in Pittsburgh may not recognize local forms, or may have different impressions of them, which shows the mismatch between perception and production. While the A-curve distribution pattern is reliably present, as we have seen, our perceptions are never so reliable, or at least not consistently guided by the facts of actual usage.

How populations of speakers differ in their actual speech production comes down to the relative frequency of variants for every linguistic feature. Features are not used categorically, even though we might be tempted to think so. To think of variants as categorical or binary is really just a matter of how they are counted; variants can always be categorized as part of a wide range of variation. It should not really come as a surprise that Southerners mostly say *y'all* but also say *you* and *you guys*, or that Canadians sometimes say [zi] and Americans sometimes say [həus]. But the public will continue to think in those categorical terms, because of the prevalence of A-curves with Distribution Type A that offer such a perceptual advantage to their top-ranked features. We linguists can do better if we use what we know about the consistent A-curve pattern in speech. We should pay attention to the 80/20 Rule, and focus on the 10 percent or 20 percent of variants that account for the great majority of actual usage. In this experiment, that 10 percent or 20 percent is a number ranging from 4 to about 40, depending on the item, a much larger number of variants than we linguists have usually considered in the past. We should not forget that there are also a great many variants that occur at low frequencies: since A-curves always have long tails, our descriptions of speech cannot afford to ignore the 20 percent of tokens that provide the greatest part of the typological richness of a language. A great many more variants are possible than we have been accustomed to recognize. These facts have strong implications for some of our traditional practices, like the use of binary judgments for whether or not some feature variant (whether lexical, phonological, or grammatical) belongs to a linguistic system. The classic North American sociolinguistic Class/Style model does not address these frequency patterns in any way. Even if style shifting does occur in some cases, it is simply incorrect to think that the Class/Style model captures much of the variation evident in different populations of speakers. The relative frequencies of the same variants across many different populations exist in a rich, multidimensional pattern, not in any simple linear framework. Finally, we also need to remember the distributional property of scaling, so that we do not expect our observations of speech over wide areas to apply at lower levels of scale, or vice versa. As reported above, we should agree with Horvath and Horvath (2001, 2003) that it is a basic fallacy to think that the behavior of any smaller group will necessarily be the same as the behavior of the larger group of which it forms a part, or that the behavior of any group overall will predict the behavior of its sub-populations. We are always right to think that speech varies between different populations, but we need to be careful not to allow our naïve perceptions, whether or not conditioned by conventional aesthetics, to overwhelm the patterns that we can actually observe in speech production.

Measurement of scale-free data

Having now considered ways of thinking about populations that are more flexible than the classic North American Class model, and having considered the pattern in which we can expect to find variation among those populations, we can turn to how the variation might best be measured. One of the great advantages of using complexity science in linguistics is that we can connect our field to the many others that study complex systems, and adopt principles and methods that have been successful in other disciplines. In an experiment designed to study the differences in A-curves in our linguistic data (reported in Kretzschmar, Kretzschmar, and Brockman (2013), on which this section is

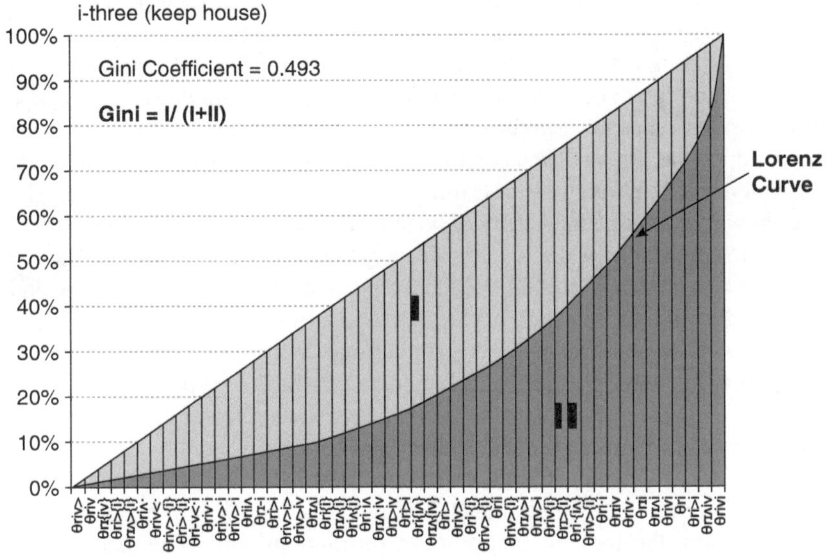

Figure 7.11 Lorenz Curve and Gini Coefficient

based), a public policy analyst and a chemical engineer collaborated with a linguist, the present author, to use the Gini Coefficient from the field of economics to model linguistic data. In economics, the so-called Lorenz Curve models the distribution of wealth in a society. The members of a population are represented on the x-axis by rank of the amount of wealth they hold, low to high in order left to right, and the percentage of the total wealth of the population is represented on the y-axis, so that a chart represents a cumulative distribution of wealth in society. In practice there is always a curve that represents the relative inequality of wealth in a population, as opposed to the hypothetical straight line of perfect equality of wealth. The Gini Coefficient gives the proportion between the actual wealth shown on the Lorenz Curve and the line of perfect equality (see Figure 7.11). As it happens, the Scandinavian countries and Western Europe tend to have low Gini Coefficients, representing greater relative equality of wealth, the USA and China have moderate Gini Coefficients, and many emerging economies have high Gini Coefficients. When the Lorenz Curve is charted, then, the deeper the curve, the higher the Gini Coefficient, and the more unequal the wealth. For our experiment we used a calculator for Gini Coefficients developed for use in yet another discipline, public health, from the East of England Public Health Observatory (www.erpho.org.uk; the ERPHO site also contains explanatory material about the Gini Coefficient and its use in public health). We modified the calculator to accept more data to accommodate our tallies, up to 300 entries. The ERPHO calculator produces the kind of chart represented in Figure 7.11, but we have used some of our own phonetic data there to generate the curve. We ranked phonetic transcription types representing /i/ in *three* (among LAMSAS speakers in the Keeping House category) by frequency, low to high in order left to right, and plotted the cumulative percentage of transcription types to form the curve. The result is a Gini Coefficient for the phonetic data that shows how steep the curve is, which represents how unequal the frequencies of phonetic transcriptions types are, just as in economics the curve shows the relative inequality of wealth. The Gini Coefficient, then, offers an index for the shape of the A-curve for any data set, where a low coefficient indicates greater relative equality of the frequencies of transcription types, and a high coefficient indicates greater inequality. In terms of the Distribution Types of A-curves discussed above, there is some range in the depth of the curve that we can quantify with the Gini Coefficient: Distribution Type A will have a high Gini Coefficient, Distribution Type B a somewhat lower coefficient, and Distribution Type C the lowest coefficient.

The Gini Coefficient describes a nonlinear distributional pattern that is different from normally distributed data. Figure 7.12 shows a normal distribution generated using the tool for this purpose on random.org (www.random.org/gaussian-distributions), with a mean of 5 and with a standard deviation of 2. Figure 7.12 shows the Gini Coefficient display, the data charted by rank and frequency (as regularly shown in this volume for A-curves), and the data charted by value and frequency from low to high (as one usually does for normal distributions). random.org generated numerical values to two decimal points, and the chart does not further categorize these values into bins but rather counts the frequency of exact values. The Gini Coefficient is low, at .142 compared to the coefficient of .493 in Figure 7.11 for variants of /i/ for speakers keeping house. Normal distributions plotted by frequency rank may appear to be curvilinear, especially as standard deviation increases, but the Gini Coefficient shows that they are not nonlinear in the same sense as distributions showing the A-curve pattern. A-curves are not just left-skewed Gaussian distributions. The mean frequency for the 85 variants of [i] in *three* illustrated in Figure 7.11 is 15 (values from Figure 7.11 are rounded to integers), but the median frequency is 3, too far from the mean for a normal distribution. The range of frequencies in that data set is 359, so a reasonable estimate of its standard deviation according to the standard method (range/4, if it were a normal distribution) would be 90; its calculated standard deviation is 44, so the estimate and the calculation are far apart. The A-curve distributional pattern is just different from the Gaussian normal distribution, and it does not make good sense to rely on means and standard deviations for such data.

While the Gini Coefficient thus offers a good way to measure the relative depth of A-curves in complex systems, our experiment found that modeling speech data effectively, the problem of fitting a curve to the data, involves two

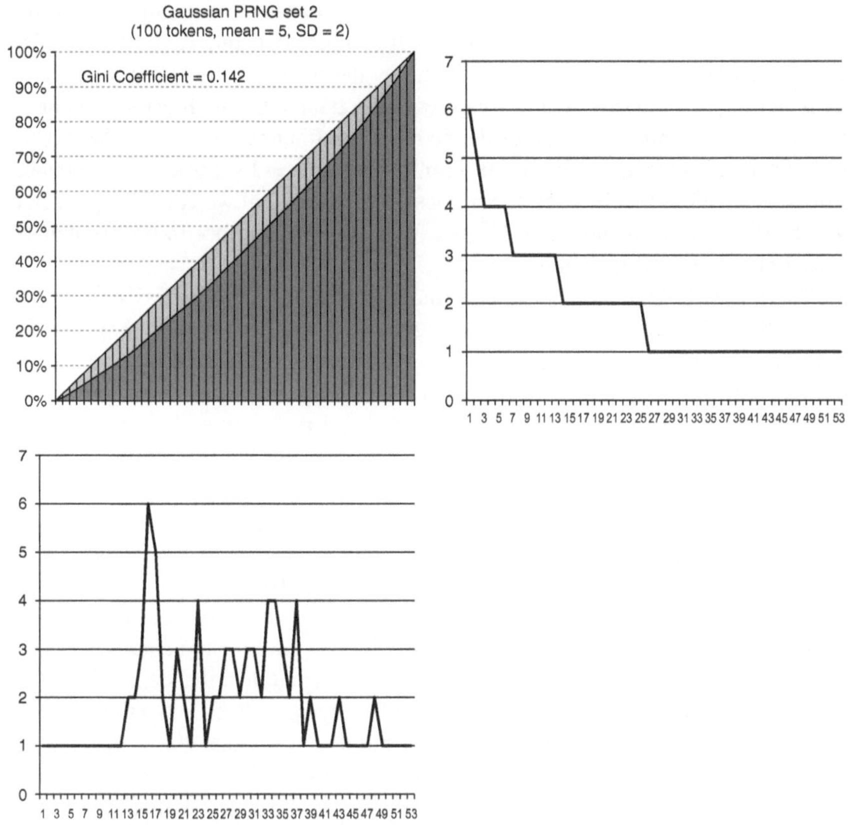

Figure 7.12 Normally distributed data set with 100 tokens (mean = 5, standard deviation = 2)

issues. The first is what statisticians call "binning," i.e. how many categories we use to divide up the continuous data. The second is sample size. We can easily conceive of the interrelation and theoretical limits for binning and sample size in a thought experiment. If we consider the problem from the binning side, we can see that a very large number of categories, say one category per token, will not show the A-curve because it yields a linear chart with a slope of zero – a horizontal line. On the other hand, frequency data sorted into just two categories also gives us a line, with a slope that depends on the difference between the two category values. Thus the outer limits for the possible number of categories are both linear, and the nonlinear A-curve can only be observed when the number of categories into which the data is sorted lies between these two extremes, between two categories and the large number of categories that represents one category for every token. The Lorenz Curve in Figure 7.11 is just such a curve between the baseline and the line of perfect equality of wealth. The Gini Coefficient, as a proportion that shows the depth of the curve, also gives us a technique to assess which treatment of a data set, which arrangement of bins and what sample size, gives us the most nonlinear chart, i.e. furthest from the two linear extremes. Since the number of bins and sample size represent limits on the Gini Coefficient, the highest coefficient that we can achieve when we try different bins and sample sizes will show us the optimum number of bins and the optimum sample size.

The first part of the modeling experiment used acoustic phonetic data in order to test the effect of the number of categories for analysis on the shape of the resulting A-curve. For a local project, acoustic phonetic vowel formants were collected from students at the University of Georgia (Figure 7.13, presented earlier as Figure 3.1). Each student carried out a fixed format elicitation task in which the cue was recorded twice. For the vowel characteristics, fundamental frequency (F0), F1, F2, and F3 of each token were measured from the display of the TF32 program implemented in CSpeech SP (Milenkovic and Read, 2000). For monophthongs, the frequencies are measured in the center of the vowel segment. Figure 7.13 presents the data from 16 speakers, half men and half women, for the vowel /u/ in five different words; there is a total of 148 observations in the data set. Presented here in grayscale from a chart originally in color, the women's tokens are represented by triangles, the men by circles, with the overall means indicated by F and M (the *pool* tokens make the separate cluster at right). For the modeling experiment, we divided the range for /u/ among these speakers into a regular grid, and counted the number of tokens in each cell of the grid. For the first measurement, we created cells 50Hz F1 by 200 Hz F2, to yield an 8×18 grid of 144 cells, as shown in Figure 7.13; of these, 59 cells had at least one token. We then changed to an 8×9 grid with 72 cells, and then to grids of 4×9, 4×6, 2×6, 2×3, and 1×3, with attendant changes in F1/F2 dimensions. For the finer grids, not all cells had tokens in them, so that not every possible state of the data was observed. For the coarser grids all cells had at least one token. The different grids thus had cells with varying numbers of tokens in them as represented in Table 7.1, four of which are charted in Figure 7.14 ranked by cell frequency in order to show the depth of the curves. Every grid yielded a different A-curve for the same data, and we calculated the Gini Coefficient for each one. For this dataset, the most nonlinear curve was the grid with 24 cells, 21 of which were occupied. Table 7.1 shows the Gini Coefficients across the seven grids that were applied to the /u/ tokens. Adding more categories, i.e. more cells with tokens in them, progressively decreased the Gini Coefficient from the optimum number of 21 categories. Similarly, using a grid with fewer cells also decreased the Gini Coefficients from the optimum number, dramatically so for the smallest numbers of categories in

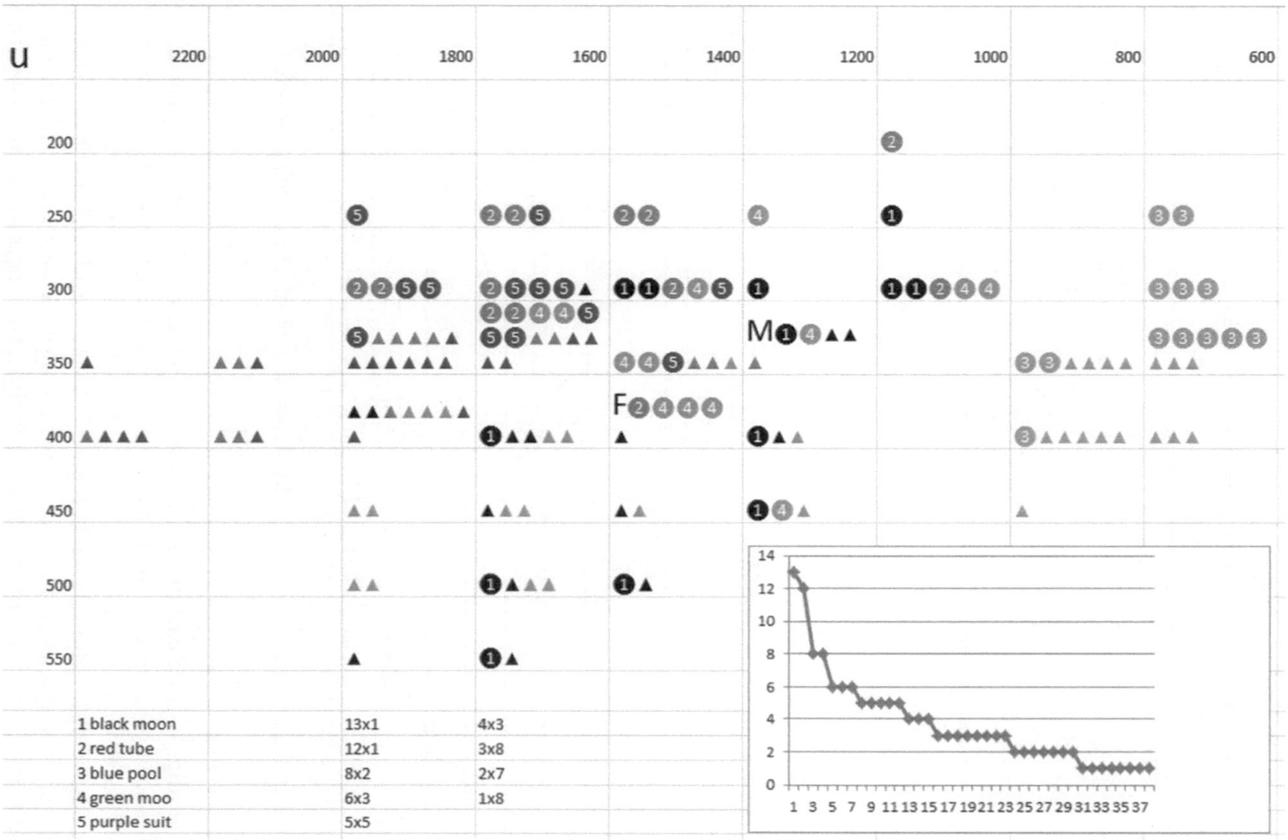

Figure 7.13 Distribution of UGA /u/ realizations

Table 7.1 *Counts and coefficients for 148 tokens in different categories*

State space	Categories occupied	Tokens/type	Gini Coefficient
144	59	2.44	0.363
72	38	3.79	0.381
36	29	5.1	0.393
24	21	7.04	0.477
12	12	12.33	0.377
6	6	24.67	0.219
3	3	49.33	0.159

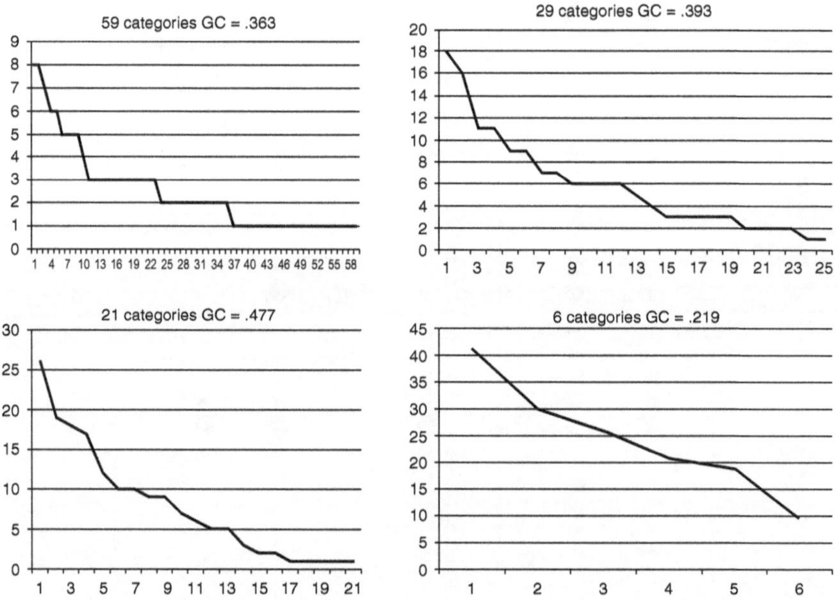

Figure 7.14 Four A-curves for different numbers of categories for one data set

single digits. Three categories is the lowest number that can show any curve at all (since two categories produce a line, as in the thought experiment), and the Gini Coefficient for three categories is .159, not unlike the Gini Coefficient for the normal distribution shown in Figure 7.12. It is clearly the case that a small number of categories will have a strong limiting effect on the Gini Coefficient. The Gini Coefficient appears to be a fairly robust measure from about ten categories on up to as many as were included in the modeling experiment, 59. The thought experiment predicts, however, that there will be a number of

Table 7.2 *American English vowels*

Vowel	Word Elicited	Simplified?
i	three	
ɪ	six	yes
e	eight	yes
ɛ	ten	
æ	half	yes
ɑ	watch	
ɔ	fog	yes
o	sofa	
ʌ	one	yes
ʊ	good	
u	two	yes
ai	five	

categories at the high end, as the tokens-per-type ratio falls towards 1. At this end of the spectrum, the Gini Coefficient will decline rapidly or become erratic, just as we see at the low end of the number of categories where the nonlinear pattern has been changed by the number of categories to become nearly linear and thus could be mistaken for a Gaussian normal distribution. Binning, then, does affect the measurement of the nonlinear distribution.

The effect of sample size on A-curves was the object in the second part of the modeling experiment. Table 7.2 shows a single word for each vowel in the American English vowel system, from the words available in the phonetically transcribed portion of the data from LAMSAS. Each data file was processed to remove inappropriate responses, and to level consonantal differences. The fine IPA transcriptions used for the Atlas allow for a great many types of realization of each vowel, once we consider all the diacritical marks that could be applied plus possibilities for diphthongization. All of the unique strings representing the vowel were tallied and ranked by frequency. For example, in Table 7.3, the tally of the different realizations of /æ/ in *half* has 88 different realizations in full phonetic transcriptions ({} indicate weakly realized segments). For a subset of six vowels, we also processed the data set with "simplified phonetics" as available on the Atlas Web site, which removes the diacritical marks from the data and thus significantly decreases the possible number of categories of realization. The number of transcriptions from simplified phonetics for /æ/ in *half* is 19, a much smaller number. While there are nominally 1,162 speakers in the LAMSAS, there was some missing data for each vowel, and multiple responses from the same speaker were sometimes recorded, so that in Table 7.4, for /i/ in the word *three*, we had 1,291 tokens; we averaged 1,173 tokens per vowel across the vowel system. We then found sociolinguistic categories tracked for the Atlas that would represent different sizes of subsamples. Men

Table 7.3 *Variant transcriptions for /æ/ in* half

Full phonetics (88 types)				Simplified Phonetics (19 types)	
æ{ə}	280	æɨ^	2	æ{ə}	449
æ	216	æ^ɨ	2	æ	364
æ·{ə}	78	æ^ɨ^	2	æ{ɨ}	116
æ·	69	ɑ	2	æ{ε}	101
æ{ε}	47	εɨ	2	æɨ	48
æ{ɨ˅}	42	ɑ:	1	æε	22
æ^{ε}	37	ɑ^<{ə}	1	a{ə}	11
æ{ɨ}	35	a^>{ə}	1	ɑ	4
æɨ	32	a>·{ə}	1	a	3
æ^	19	ɒɒ̥^	1	εɨ	3
æ^·	16	ɑ>ɒ̥^<	1	æ:	2
æ>	16	æ:	1	ɒ	2
æ˅>	16	æ^:	1	ɑ:	1
æ˅{ə}	15	æ˅{ə}	1	ɑ{ə}	1
æ{ə^<}	13	ǎ{ə}	1	ɑɒ	1
æ^ε	13	æ^{ə^<}	1	ɒɒ	1
æ^·{ə}	12	æ^{ε>}	1	ææ	1
æ^{ə}	11	æ{ɨ^}	1	æə	1
æ>{ə}	10	æ>{ɨ˅}	1	eɨ	1
æ·{ɨ˅}	10	æ˅{ɨ}	1		
æ˅>{ə}	8	æ˅>{ɨ}	1		
æ˅	8	æ·{ε<}	1		
æ^{ɨ}	7	æ·{ə˅}	1		
æ{ə^}	6	æ>·{ə}	1		
a^{ə}	5	æ<·{ɨ}	1		
æ{ε>}	5	æ>·{ɨ}	1		
æ·{ε}	5	æ^·{ɨ^}	1		
æ·{ə^<}	5	æ^·ε	1		
a^·{ə}	4	æ^·ə	1		
æ^{ɨ˅}	4	æ^·ε^	1		
æ·{ə^}	4	æ˅·	1		
æ·ε	4	æ˅>·	1		
æ^·{ε}	3	æ·ɨ^	1		
æ·{ɨ}	3	æ>·ɨ	1		
æ·ɨ	3	æ^·ɨ^<	1		
ɒ̥<·	2	ææ̜^	1		
ɑ>·	2	æε	1		
æ>{ɨ}	2	æ>ɨ^	1		
æ^{ɨ^}	2	æ^ɨ˅	1		
æ^·{ə^<}	2	a^	1		
æ^<·{ε>}	2	ă^	1		
æ^·{ɨ˅}	2	a^>	1		
æ^<·	2	ε·ɨ	1		
æε	2	e˅ɨ	1		

Table 7.4 *Data from /i/ in* three, *by categories of different sizes*

Category	Speakers	% speakers	Tokens	Types	% types	% types/ speakers	Tokens/ types	Gini
All	1162	100	1291	85	100	1	15.19	0.782
Men	802	0.69	899	72	0.85	1.23	12.49	0.775
Type 1	582	0.5	635	66	0.78	1.55	9.62	0.695
Farmers	568	0.49	617	59	0.69	1.42	10.46	0.736
Type 2	439	0.38	506	59	0.69	1.84	8.58	0.722
Women	360	0.31	392	61	0.72	2.32	6.43	0.664
KeepHouse	278	0.24	305	52	0.61	2.56	5.87	0.64
Type 3	138	0.12	150	35	0.41	3.47	4.29	0.632
Managers	96	0.08	122	26	0.31	3.7	4.69	0.646
Professionals	47	0.04	54	15	0.18	4.36	3.6	0.558
New York City	25	0.02	25	7	0.08	3.83	3.57	0.457
Philadelphia	11	0.01	11	4	0.05	4.97	2.75	0.341

represent a subsample of 69 percent of LAMSAS speakers, while women are a subsample of 31 percent. The three "Type" classifications, representing educational level and social involvement, represent subsamples of 50 percent, 38 percent, and 12 percent. Four occupational categories cover additional levels: Farmers, 49 percent; people who Keep House, 24 percent (this is the category of speakers represented in Figure 7.11); Managers, 8 percent; and Professionals, 4 percent. At the lowest scale of subsampling, people interviewed in New York City represent 2 percent of the overall sample, and people interviewed in Philadelphia represent 1 percent. In Table 7.4 the subsamples are sorted in order of decreasing proportion of the overall sample. The table shows the number and proportion of speakers in each subsample, the number of tokens elicited, and the number and proportion of types of unique vowel strings found in each subsample. In all, this part of the experiment included 216 data sets: twelve vowels with full phonetics and six vowels with simplified phonetics, for the sample overall and for eleven subsamples graded in size from 69 percent down to 1 percent.

Our initial finding in this part of the experiment leaves no doubt about the relevance of complex systems to language variation: the A-curve pattern appeared over 99 percent of the time across our 216 tables, lacking just two Philadelphia cases in which the small number of speakers produced fewer than the three variants needed to draw a curve. The average values are also quite revealing (Table 7.5). The overall average Gini Coefficient of .758 matches the theoretical value of a continuous exponential function fulfilling the 80/20 Rule, as calculated by our engineering colleague, which is a Gini Coefficient of .752. The averages for the full phonetic transcriptions show that the Gini

Table 7.5 *Averages across all vowels*

Category	Speakers	% speakers	Tokens	Types	% types	% types/ speakers	Tokens/ types	Gini
All	1162	100	1172.67	122.17	100.00	1	13.43	0.758
Men	802	0.69	814.5	98.83	0.82	1.19	11.18	0.723
Type 1	582	0.5	575.58	89	0.74	1.49	8.72	0.692
Farmers	563	0.49	569.83	79.75	0.66	1.35	9.91	0.707
Type 2	439	0.38	452.92	72.5	0.6	1.59	8.67	0.677
Women	360	0.31	358.17	73.33	0.61	1.96	6.92	0.642
KeepHouse	278	0.24	274.75	62.92	0.53	2.22	5.92	0.626
Type 3	138	0.12	144.17	43.67	0.38	3.21	4.05	0.565
Managers	96	0.08	106.08	35.67	0.33	3.98	3.49	0.512
Professionals	47	0.04	51.25	21.42	0.13	4.56	3.06	0.43
New York City	25	0.02	25.33	9	8	3.84	3.96	0.411
Philadelphia	11	0.01	11.08	5.5	0.05	5.44	2.94	0.280

Coefficient does indeed decline steadily as sample sizes become smaller. The Gini Coefficient initially stays quite robust as the samples get smaller; however, as the subsamples get quite small, the decline in Gini Coefficient accelerates. We see here the same effect that we encountered in the first part of the experiment, that Gini Coefficients fall off sharply at the low end, now for sample size just as they did for binning.

One unexpected and important finding of the modeling experiment is the relationship between sample size and the number of types observed (Figure 7.15). If we compare the percentage of types found overall to the percentage of speakers in any subsample, we can create a Type/Speaker index by dividing the percentage of types by the percentage of speakers (column 7 of Table 7.5). This index begins at 1, since 100 percent of the types occur among 100 percent of the speakers, but then the Type/Speaker index increases with every smaller subsample. The effect is moderate in the larger subsamples, but at Keep House (with 24 percent of LAMSAS speakers), where the decline for Gini Coefficients becomes sharper, we see a substantial increase in the Type/Speaker index. There is a dip in the trend in New York City, but the index there is still much higher than we would otherwise expect, nearly four times as high. The smaller samples retain much more of the variation in types found in the overall sample than we would expect them to, over five times as much for the smallest sample in Philadelphia. We might explain this observation with reference to the inherent multidimensionality of the complex system of speech: since there are many local and social groups in a large sample, each of which may have different top-ranked variants for any feature, and the same multiple dimensions are still in play for small samples, then the small samples will preserve more variation than the number of speakers would suggest as a consequence of the dimensionality

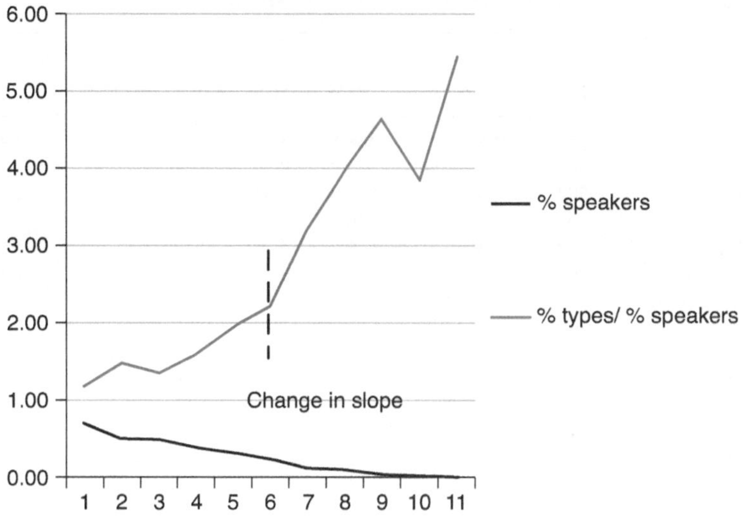

Figure 7.15 Type/Speaker index

of speech. Speech does not become less complex just because we talk to fewer people. This trend multiplies the limiting effects of smaller samples on the Gini Coefficient by reducing the type/token ratio for them. If we want to use small samples to study language data, the nature of the complex system of speech is working against us. Here is proof, if we still need it, that no single speaker or small group of speakers can be representative of a large speech community.

The modeling experiment gave us one more interesting result, from the smaller number of categories arising from simplified phonetics. Here the averages only cover six vowels, but the change in Gini Coefficients is very strong, much higher than the values from the full phonetics categories (shown at right in Table 7.5) except for New York City and Philadelphia, the two smallest subsamples. For this data, the number of categories derived from simplified phonetics is clearly closer to the optimum than the larger number derived from full phonetics. The Gini Coefficients remain above .700 all the way down to people who Keep House, the threshold for the small sample size effect, and remain above .600 for the two categories below that in size. Why? The number of simplified phonetics types overall only averages 23, compared to an average of 122 with full phonetics. This means that every extra retained variant in the small samples has proportionately more weight. More descriptive phonetic detail is not necessarily better.

Our results in the modeling experiment give a strong indication of how the scaling nonlinear distributions of variation data work in practice. Sample size and the extent of categorization do indeed affect the shape of the A-curve in an

Table 7.6 *Average over all vowels, simplified phonetics*

Category	Speakers	% speakers	Tokens	Types	% types	% types/ speakers	Tokens/ types	Gini
All	1162	100	1188.83	23.17	100	1	74.89	0.795
Men	802	0.69	825.33	19.5	0.87	1.25	57.88	0.771
Type 1	582	0.5	584.5	17.33	0.73	1.46	52.82	0.722
Farmers	568	0.49	576.67	16.33	0.7	1.43	53.71	0.731
Type 2	439	0.38	456	16.33	0.72	1.9	38.86	0.728
Women	360	0.31	363.33	16.17	0.72	2.32	30.69	0.722
KeepHouse	278	0.24	279.33	15.17	0.68	2.85	24.57	0.708
Type 3	138	0.12	148.17	11.33	0.58	4.82	15.15	0.636
Managers	96	0.08	107.33	10	0.52	6.4	11.91	0.603
Professionals	47	0.04	52.33	7.5	0.38	9.34	8.07	0.524
New York City	25	0.02	25.17	3	0.16	7.85	9.22	0.414
Philadelphia	11	0.01	11	2.17	0.11	11.72	5.81	0.275

interrelated way. Gini Coefficients do work well to assess the differences in A-curves that arise from these factors. Unexpectedly, small samples retained more variants than one would expect given the proportion of the subsample out of the sample overall, and a great deal more variation than expected in the smallest samples. Our results do not just come from the Atlas field workers or survey method, because we also found scale-free, nonlinear patterns in the acoustic phonetic data we collected locally from students at Georgia. The modeling experiment shows us clearly that working with only a small number of types (bins), and/or with only a small sample, is highly likely to give misleading results. We cannot use small indexes of the kind frequently encountered in classic North American sociolinguistics, or the small samples also frequently found there, to get, in Labov's words, "a clear view" of language behavior.

Complex systems for sociolinguists

Now, with this more detailed picture of the patterns that emerge from nonlinear distribution and scaling, we are ready to consider how such complexity is relevant for sociolinguists. Let us begin by recalling the five basic principles of all complex systems (Table 7.7). Each of these principles maps easily to what we know happens every day in our speech communities. The one exception may be the idea of random interactions. Of course we cannot deploy words or phonetic segments randomly and still make sense, but speakers do use the *variants* of any linguistic feature randomly, at least before the complex system has operated on them. The fact that variants of features interact randomly is demonstrated by the establishment of new language varieties in new settlements (as discussed

Table 7.7 *Principles of complex systems applied to speech*

1 continuing dynamic activity in the system = conversations and writing	
2 random interaction of large numbers of components = random variant realizations of features	
3 exchange of information with feedback = comparison of components used by different speakers in different settings	
4 reinforcement of behaviors = changing likelihood of components in future uses	
5 emergence of stable patterns without central control = regional/social/textual varieties	

above). Moreover, the diversification of a variety into different subvarieties, by gender, type, occupation, and locality as the charts from [i] in *three* show us in Figure 7.16, is a fundamental process in the operation of the complex system. We can also see in Figure 7.16 how the charts for the two smallest categories appear to be more linear. While these domains and categories are still the ones used for the modeling experiment, the scaling behavior of the complex system is literally unlimited. The underlying randomness and indeterminacy of the components in the complex system keeps our languages and varieties from freezing up, and allows them to change continuously to address new circumstances across space, society, and time. Of course these scales interact, as for women in different occupations and locations and as members of the overall pattern, or as for Managers of different genders and educational levels in different places. Nonetheless, the scaling property of the complex system is what allows us to preserve the wide range of different patterns in any language that we never fail to observe. We humans are smarter than molecules, cells, or ants as they interact in their own complex systems, and so we can command a great many more behaviors in language as they emerge in all of the different regional, social, and textual settings for speech. In this way, complex systems are fundamental to sociolinguistics in particular among all the ways that we study language.

It is also clearly the case, however, that classic North American sociolinguistics never addresses these distributions. For one thing, the preoccupation of North American sociolinguistics with grammars has led it to focus on a perceptual quality of language variation, not on observations of speech production per se. Even Labov's founding insight is framed as coexistent systems, parallel grammars, not as anything that observes and analyzes distributional patterns. The Class model does include nine socioeconomic levels, a set of parallel grammars rather than just one grammar for the language overall, a movement towards the infinite multidimensionality that we recognize in complex systems. However, these categories are often regrouped into a smaller number of categories (e.g., into six categories in Figure 7.1), and Labov's focus on the working class as the locus of change in both New York and Philadelphia backs away from a commitment to parallel grammars across all of society. Labov's practice

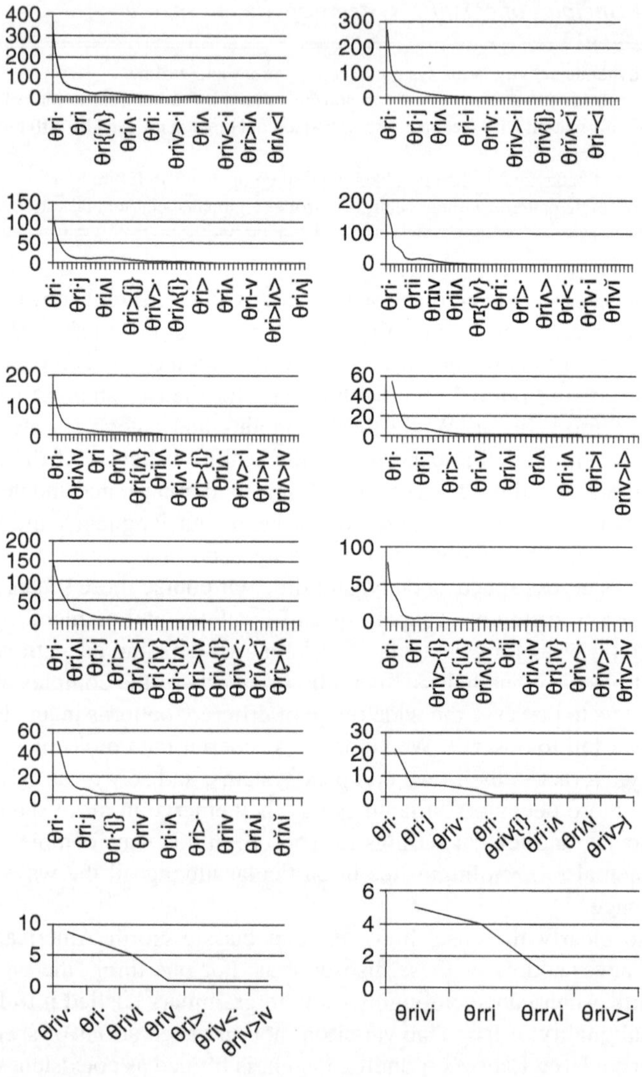

Figure 7.16 Twelve A-curves for subsamples of /i/ in *three*

of paring away populations who do not conform to his working-class grammar, such as ethnic speakers, also reduces the dimensionality of his model. Complexity science tells us that nine levels are not enough, even if North American sociolinguists always worked with all nine. Labov's Style model also fails to grasp what actually motivates it, the fact that each style is really its own text type. The measurement experiment has shown us that the appearance of a linear relationship between the styles may well arise as an artifact of the small number of categories for analysis, which can make the underlying nonlinear frequency distribution appear to be more like a normal distribution. The foundations for North American sociolinguistics render it incapable of handling the output of the complex system of speech in relative frequencies of all linguistic variants of features across all populations of speakers. It does not diminish Labov's achievement, his launch of quantitative methods and socially differentiated analysis for linguistics, to propose now that we are further down the path that we can refine sociolinguistic methods to do a better job of accounting for what we now understand about complex systems.

The first sort of recommendation addresses the "parallel" part of parallel grammars: complex systems suggests more deliberate handling of populations and sampling issues. The second kind of recommendation addresses the "grammars" part of parallel grammars, how we make categories to study the variation we see, and our choice of grammatical models.

First, sociolinguists would do well to be more exact in naming the populations of speakers for any study. Complex systems tell us that *any* sufficiently large group will form its own frequency distributions of features, and so we are at high risk of bias from measuring smaller groups than we intend, or alternatively from measuring larger groups than we intend, if we do not explicitly identify who is eligible to be one of our speakers. So, for example, the classic sociolinguistic snowball (or "friend of a friend") method of acquiring speakers in a community is *likely* to introduce bias since the people are known to each other, *prima facie* evidence that they are involved in a social network – so that we will be measuring speech just in the social network instead of speech in the larger community. Sociolinguists *may* choose to study any group of speakers, and can validate the choice of a group by asking the speakers if they are members of it, but at the same time sociolinguists *must* identify exactly what group they intend to study.

A corollary to the practice of exact description of a target group of speakers is for sociolinguists to take an interest in the other levels of scale in which their speakers participate. It is no longer good enough just to assume that upper-class speakers use the "standard" language. Language standards, in fact, are perceptual and institutional creations, not an outcome of the complex system of speech, as discussed in several earlier chapters of this volume. Standards may or may not institutionalize the most frequent variants for a national variety, and speakers today generally try to enact them, or do not, mostly on the basis of their commitment to educational institutions in which the standards are transmitted. James and Lesley Milroy's long-running book, *Authority in Language* (3rd edn., 1998), does a fine job of separating the "complaint" tradition from the actual circumstances of language in use, and what they say accords well with a foundation in complex systems. The Milroys have also led the way in their comparison of target groups of socially networked speakers in Belfast (L. Milroy 1987; J. Milroy 1992), in which they compared the variants found in different neighborhoods to each other, and also to the findings of their "doorstep" survey across the city. This kind of multiple comparison, the "correlational studies" that Labov rejected as of little interest, allows the sociolinguist not just to say what variants may be more common in a target group of speakers, but also to say whether these variants are candidates for perceptual salience as discussed earlier in this chapter. "Standard English" is not one pole in a linear dimension of formality, and neither is it the property of upper-class speakers. This means that sociolinguists should use evidence gathered from multiple groups of speakers for comparison to the evidence from a target group of speakers, in order to tell how the frequency profiles of the target group are embedded in a multidimensional matrix of speech. That said, because of the scaling property of the complex system, it will always be possible to contrast the frequency profiles of any smaller group against those of the relevant larger group, say the speech of a neighborhood against the speech of the larger city in which it is found.

Following from proper identification of the group chosen for study, sociolinguists should use randomized sampling techniques to avoid unintentional bias from groups other than the one under study. Tagliamonte has written of "the often extreme difficulty in finding the subsection of the population that you wish to study" and, given that problem, "random selection of participants with uniform probability over the entire population is, in fact, not very useful" (2006: 22). It is certainly true that one should not create a random sample of an entire city if one only intends to study a single neighborhood. However, having decided on one neighborhood, or another demographic segment, it is important to use randomized sampling to select participants. All members of the study group will also belong to other groups as well, and researchers need to avoid unintentional bias from those other connections. The only way around this problem is randomized sampling, in which every member of any community we identify has a chance to be interviewed in the sample. For the data used frequently in this volume, I have demonstrated (Kretzschmar et al. 1993: 16–19) that our LAMSAS survey does fit within the parameters of modern randomized survey research for its target population, with the same confidence interval as commonly adopted for contemporary election polling. Sociolinguists should be able to make the same claim. Tagliamonte proposes stratified random sampling as a reasonable alternative in study design (2006: 23–24), especially with use of principles of random selection to identify speakers for each quota. It is true that one cannot make a strictly random sample of human beings from any large population. People are not like light bulbs, bottles of beer, or other products of manufacturing processes that can be identified comprehensively and sampled precisely. Still, we can agree with Tagliamonte that stratified random sampling is a good alternative in practical terms.

The LAMSAS survey is quite large, but sociolinguists need not plan work on that scale; the experiment with Gini Coefficients suggests that samples of at least 30 observations would be a good idea. Researchers should strive for data sets with 250–300 observations, since in both parts of the modeling experiment the Gini Coefficients declined at a more rapid rate below that number. However, Chambers (2009: 44–45) and Tagliamonte both quote David Sankoff to the effect that "ongoing work in sociolinguistics found that relatively small samples – samples too small to be technically representative – were sufficient to account for language variation in large cities" (Tagliamonte 2006: 23). Milroy and Gordon, too, quote Sankoff, noting that "samples for linguistic studies tend to be much smaller than those found in other types of surveys" (2006: 28–30). Tagliamonte recommends "reaching the right people" (2006: 28) for sociolinguistic purposes, and that "it is better to design your sample to be smaller . . . the size of the sample must necessarily be balanced with the available time and resources for data handling" (2006: 33). These arguments for small samples are simply wrong, if there are too few of "the right people." Tagliamonte is certainly right that sample size must be balanced against time and resources, but still the sample must reach the critical size identified in the Gini Coefficient experiment in order to provide a clear view of the distributional pattern of variation. The axiom of categoricity may have led sociolinguists to think that very small samples could be acceptable, and the popularity of sociolinguists has come in part from the idea that anybody can collect data from a small sample, but the dynamics of measurement in the complex system will not allow it.

Some sociolinguists will say, with good reason, that random sampling is unlikely to obtain the kind of casual speech we can get from the snowball method, which prioritizes casual conversation among friends and acquaintances. To clarify the grounds of this problem, we should remember that conversations using race- or gender-appropriate interviewers, or friends rather than strangers, or groups rather than single speakers, all enact different situations for speech, different text types. From the point of view of complexity, the emphasis in North American sociolinguistics on the vernacular has actually been a preference for one text type, casual conversation, out of the unlimited number of settings for speech. Thus, sociolinguists can continue to recognize differences of style and recognize the problem of the observer's paradox, but now as problems concerning text types. Sociolinguists should become even more explicit about the situations for speech that we create in order to observe language variation, and about the groups we do that for. We should choose what we want, and then accurately describe what we have. So, for instance in the

Roswell Voices language and life project (Kretzschmar et al. 2007), we have encouraged people to speak freely about topics like the role in daily life of local churches, schools, and institutions, and speakers' relationship with the big city of Atlanta. Such a guided interview protocol gives us interviews with lots of narrative passages. This method has generated data for us that we consider to be quite comparable as text types from speaker to speaker. Sociolinguists cannot, however, continue to prefer casual speech, the vernacular, as the best or most basic variety. Casual speech is just another text type among many. Speakers always have to negotiate their performance in many different situations, and none of them is more basic than any other.

To turn now to grammars, sociolinguists need to account for the nonlinear frequency pattern that emerges from the complex system in order to make abstractions about language as a hierarchical system, as in phonology and syntax. The more that sociolinguists try to use binary distinctions or create indices which fit the data into a small number of categories, as Chambers implied by recommending "regularization" of variation, and as Chambers and Trudgill (1998: 51ff.) have taught in their textbook on language variation, the more we are at risk of misrepresenting the underlying nonlinear pattern. An index is an abstraction, and integers like 1, 2, 3 in the index are transformations of linguistic relationships as between (in Chambers and Trudgill 1998) [t], [tʔ], and [ʔ]. The realization [tʔ] is not "intermediate" between the other two articulations in the same way that 2 is intermediate between 1 and 3, and one cannot say that the articulations can be "averaged" in the same way that integers can. Even if no index is made and realizations are grouped into a small number of categories, the modeling experiment clearly showed that numbers of categories for variants in the single digits, like typical sociolinguistic comparisons, force the distribution to become more linear than it would be if more categories were employed. This may make the data more susceptible to analysis with commonly used statistics like linear regression, but such data transformations distort the distributional pattern.

We can return to statistics to illustrate this problem. When different independent variables are created for a Varbrul run, the program uses stepwise regression to select ordered factor weights that wash out the nonlinear pattern of variants that exists for each of the variables. In a complex system, we should be interested in the frequency ranking of all the variants, not just the common ones. The top-ranked variant, which is what Varbrul will pick out, should be considered in the context of the order of variants in the frequency profile. This means that no analyst should throw out the most-frequently occurring type because it occurs too often and obscures the pattern; every feature will have a massively more frequent variant as part of its nonlinear pattern. The basic problem with variable rules as applied to linguistic environments is that they are unlikely ever to treat all of the variation that actually occurs. 80 percent of the types that actually occur generate only 20 percent of the tokens observed, while just a few of the types, about a fifth of them, generate 80 percent of the tokens observed. Variable rules therefore may do well to capture the few variants that occur most often, but they will probably miss the 80 percent of the variant types that actually occur, or force the infrequent types into mixed categories that each contains a range of internal variation. A parallel problem arises with use of the Varbrul program because the program is designed to reduce the dimensionality of the data set. Using the statistical program R instead of Varbrul does not fix this problem, if the analyst chooses statistics that assume a normal distribution and reduce the dimensionality of the data. As we now understand from study of complex systems, retaining the multidimensionality of the data distribution is not a fault that blurs our focus on variables of interest, but instead a highly desirable practice.

Any decision to collapse the number of variables really comes down to the grammatical model we choose. As discussed in Chapter 4, the generative model prefers grammaticality judgments, binary distinctions, and the binary model is a terrible fit for complex systems because of the 80/20 Rule. Generative and structural grammars both seize upon the most frequent features in the nonlinear distribution and assign them canonical, if not categorical status, at the expense of the large proportion of data types that occur only infrequently. Optimality Theory is designed to address distinctions in small arrays rather than binaries, but the number of possibilities considered in such analysis is still too few to address the underlying nonlinear distribution of the data. Similarly, Tagliamonte's discussion of categorical, near categorical, and variable contexts (2006: 86–87) mistakes a grammatical model for a distributional pattern. There are no categorical contexts, so long as the analyst looks for the variability in them. The way for sociolinguists to increase the number of grammatical categories would be to use techniques from construction grammar to study form/function relations, as illustrated in Chapter 3. Construction grammar preserves the characteristics of the output of the complex system by considering all of the different words that can be used in a construction, where simple binary distinctions or small indices do not. As a part of the usage-based linguistics now prominent in language acquisition studies, construction grammar is worth considering for those other usage-based students of language, the sociolinguists.

Finally, it will not have escaped the notice of the careful reader that the results of the modeling experiment presented in this chapter offer a challenge to a very large number of sociolinguistic studies over the past fifty years, all of those that used small samples and a small number of categories in their analyses. However, it would make no more sense to discard all of that sociolinguistic labor and insight than it would to throw away the data and insights of the Atlas project, which also was developed before complexity science was known. Sociolinguists did hear and record actual speech from all of the speakers in their interviews.

Just as it has been possible to reanalyze Atlas data with new principles from complexity science, so, too, can we look again at our sociolinguistic data sets. In order to make that possible, it would be highly desirable for those data sets to be made public so that everyone in the field has the chance to look at nonlinear distributions and scaling in the raw data (for recommendations about such publication in an ethical, informed manner, see Kretzschmar et al. 2006). Even without the raw data, it will often be possible to revisit sociolinguistic articles and books to look for ways to solve problems with data handling and analysis that the original investigators, of course, had no way to anticipate. At the very least, we can learn from the presentation of sociolinguistic data as it stands, and refrain from just accepting generalizations made from the data that violate the distributional and scaling properties of complex systems. Such generalizations might still turn out to be good ones, if we reanalyzed the data without applying the axiom of categoricity. We should not throw the baby out with the bathwater by rejecting out of hand the findings of traditional sociolinguistic studies.

As we have seen, the new science of complex systems is a very good fit indeed for sociolinguistics. It matches up well with the founding insights of the field. Complexity science predicts the kind of patterns that occur in speech, and also predicts the unlimited variety of possible speech communities that emerge and coexist at different scales and continue to change over time – just the broad range of varieties that attracted sociolinguists in the first place. The central value of complexity science for sociolinguists is thus to define the relationship between what people actually say and write, and the generalizations we wish to make from speech, whether or not we choose to make formal grammars, and whether our generalizations describe regional, social, or textual varieties. Some sociolinguists, no doubt, will prefer to retain the axiom of categoricity, and to continue to create more formal analyses that stay closer to form/function relations in grammars. In so doing, they might well improve their discussion of grammars with the recommendations of Chapter 4. A knowledge of complex systems enriches what sociolinguists have achieved by allowing us to see the earlier work in a new light. When we drop the axiom of categoricity, and when we take due account of the small samples and indices used in the past, we can reinterpret the data that has already been collected so as to make generalizations from it more appropriate to its underlying nonlinear distribution and scaling. And going forward, complexity science suggests that we can continue to do what we do in sociolinguistics, with more confidence that our fieldwork and analyses are true to our speakers and what they tell us.

8 Postmodernism and complex systems

In 1996 Alan Sokal published an article in the journal *Social Text* entitled "Transgressing the Boundaries: Toward a Transformative Hermeneutics of Quantum Gravity." The article was a hoax, "a parody article crammed with nonsensical, but unfortunately authentic, quotations about physics and mathematics by prominent French and American intellectuals" (Sokal and Bricmont 1998: ix). The book that followed the article, Sokal and Bricmont (1998), included yet more "nonsense" about math and science from writers like Lacan, Kristeva, and Baudrillard, this time with explanations why the quotations were nonsensical. In another article explaining the hoax, Sokal says that (Sokal and Bricmont 1998: 268–269):

> One of my goals here is to make a small contribution toward a dialogue on the Left between humanists and natural scientists – "two cultures" which, contrary to some optimistic pronouncements (mostly by the former group), are probably farther apart in mentality than at any time in the past fifty years . . . my concern is explicitly *political*: to combat a currently fashionable postmodernist/poststructuralist/social-constructivist discourse – and more generally a penchant for subjectivism – which is, I believe, inimical to the values and future of the Left.

Unlike the issue of Chapter 2, whether the humanities and social sciences can be subsumed by modern science and reductionism, Sokal complains about "subjectivism" among postmodern intellectuals that uses words from modern science but not the sense of the words that modern scientists would understand from them. This usage falls under what Sokal and Bricmont (1998: 1) call:

> "postmodernism": an intellectual current characterized by the more-or-less explicit rejection of the rationalist tradition of the Enlightenment, by theoretical discourses disconnected from any empirical test, and by a cognitive and cultural relativism that regards science as nothing more than a "narration," a "myth" or a social construction among many others.

Sokal's hoax, then, responded to what he believed to be inappropriate uses of modern science made by humanists, in order to combat what he saw as postmodernism.

In order to conclude this discussion of complexity science in the study of language, we should consider once more the larger question of scientific epistemology, this time from the opposite, postmodern point of view.

Fashionable nonsense

There is no question that Sokal and Bricmont have assembled a large number of cases in which humanists have used scientific terms in ways that do not make sense according to the usual interpretation of the terms within scientific contexts. They call Sokal's collection a "dossier" (Sokal and Bricmont 1998: ix), a term which in English (the book was originally written in French) carries a negative semantic prosody from its association with police reports and criminal trials. Their indictment begins with a quotation from Jacques Lacan in which Lacan compares mental disease to some non-intuitive topological objects, including a Möbius strip and a torus. Sokal asks, "Perhaps the reader is wondering what these different topological objects have to do with the structure of mental disease. Well, so are we" (Sokal and Bricmont 1998: 19). In the discussion that follows, Lacan is reported to insist that "This torus really exists and it is exactly the structure of the neurotic" (Sokal and Bricmont 1998: 20), that his use of the topological figure is not an analogy or metaphor. Further, Lacan is reported to have said that "In this space of jouissance, to take something that is bounded, closed constitutes a locus, and to speak of it constitutes a topology" (Sokal and Bricmont 1998: 20), to which Sokal and Bricmont reply (Sokal and Bricmont 1998: 20–21):

> In this sentence, Lacan has used four technical terms from mathematical analysis (*space, bounded, closed, topology*) but without paying attention to their *meaning*; the sentence is meaningless from a mathematical point of view. Furthermore – and most importantly – Lacan never explains the relevance of these mathematical concepts for psychoanalysis.

This, then, is the plan of Sokal and Bricmont (1998), to produce quotations like Lacan's and to explain the scientific meaning of each term in order to show that Lacan and other humanists did not use them as the scientists did. The summary judgment they provide for Lacan makes a larger point (Sokal and Bricmont 1998: 37):

> The most striking aspect of Lacan and his disciples is probably their attitude towards science, and the extreme privilege they accord to "theory" (in actual fact, to formalism and wordplay) at the expense of observations and experiments. After all, psychoanalysis, assuming that it has a scientific basis, is a rather young science. Before launching into vast theoretical generalizations, it might be prudent to check the empirical adequacy of at least some of its propositions.

The problem as set forth here is that Lacan has run off with terms of mathematics and used them for his own purposes. Lacan's discussion of topology indeed does not carry over the sense of topology from mathematics, and to a mathematician it therefore appears meaningless. On the other hand, Lacan and his disciples found value in this kind of language, disconnected as it was from its origins. A question not addressed by Sokal and Bricmont is *why*, for what purpose Lacan and others ran off with terms from mathematics and science.

Alan Sokal attributes part of the blame for this disconnection between science and the humanities on the teaching of science (Sokal and Bricmont 1998: 277):

> The teaching of mathematics and science *is* often authoritarian; and this is antithetical not only to the principles of radical/democratic pedagogy but to the principles of science itself. No wonder most Americans can't distinguish between science and pseudoscience: their science teachers have never given them any rational grounds for doing so.

This argument means little in cases like Lacan or Kristeva; as Sokal and Bricmont report (Sokal and Bricmont 1998: 278), university graduates are half as likely to believe in creationism, for example, as high-school graduates, and we can hardly attribute faulty low-level education as the cause of Lacan's and other postmodernists' repurposing of scientific terminology. Problems with science education have more to do with reactions to science in public policy, notably in the recent political treatment given by both the Left (as in Al Gore's *An Inconvenient Truth*) and the Right (as in the common rejection of climate change science by Republican politicians). Lacan, Kristeva, and others like them, however, do not fail to understand scientific principles (although that is possible) as much as they use the language of science to accomplish different ends.

Another major target of Sokal and Bricmont (1998) is what they call "epistemic relativism," the idea that "the truth or falsity of a statement is relative to an individual or to a social group" (Sokal and Bricmont 1998: 51), and this idea does come closer to postmodernism. As Sokal and Bricmont point out (Sokal and Bricmont 1998: 52):

> There is no doubt that the relativist attitude is at odds with scientists' idea of their own practice. While scientists try, as best they can, to obtain an objective view of (certain aspects of) the world [with allowance in a footnote for "nuances" of the word *objective*, as in doctrines like realism, conventionalism, and positivism], relativist thinkers tell them that they are wasting their time and that such an enterprise is, in principle, an illusion. We are thus dealing with a fundamental conflict.

Sokal and Bricmont review notions like solipsism and skepticism, and conclude that no statement about the world can be proven completely, but such statements can be proven beyond a reasonable doubt (Sokal and Bricmont 1998: 60). Epistemic relativism, then, they assign to the area of unreasonable doubt. Thus, in their discussion of the "strong program" in the sociology of science (Sokal and Bricmont 1998: 85–99), they conclude by saying that

> It is not enough to study the alliances or power relationships between scientists, important though they may be. What appears to a sociologist as a pure power game may in fact be motivated by perfectly rational considerations which, however, can be understood as such only through a detailed understanding of the scientific theories and experiments.

Sokal and Bricmont do not want to allow assertions of social power to get in the way of objective scientific reasoning that may be available. This argument also applies to the work of Thomas Kuhn. They will not complain if Kuhn says that "[scientific] evidence available at the time was weaker than is generally thought and that non-scientific considerations played a role"; they do, however, reject the idea that "changes of paradigm are due principally to non-empirical factors" (Sokal and Bricmont 1998: 75). The rejection comes down to what Sokal wrote about his hoax (Sokal and Bricmont 1998: 269):

> But why did I do it? I confess that I'm an unabashed Old Leftist who never quite understood how deconstruction was supposed to help the working class. And I'm a stodgy old scientist who believes, naively, that there exists an external world, that there exist objective truths about that world, and that my job is to discover some of them. (If science were merely a negotiation of social conventions about what is agreed to be "true," why would I bother devoting a huge fraction of my all-too-short life to it? I don't aspire to be the Emily Post of quantum field theory.)

While there are admitted problems with what might be "objective," Sokal insists on the objectivity of science as the anchor that saves us from an endless drift of negotiation of the etiquette of truth.

Stanley Aronowitz, one of the editorial team at *Social Text*, does not put things in quite the same way. In a reply to the Sokal hoax article, he attacks the notion of objectivity (Aronowitz 1997):

> So the issue is not whether reality exists, but whether knowledge of it is "transparent." Herein lies Sokal's confusion. He believes that reason, logic, and truth are entirely unproblematic. He has an abiding faith that through the rigorous application of scientific method nature will yield its unmediated truth. According to this doctrine there are "objective truths" since the earth revolves around the sun, gravity exists and various other laws of nature are settled matters. So Sokal never interrogates the nature of evidence or facts, and simply accepts them if they have been adduced within certain algorithms that bear the stamp of "science."

This statement is too strong, since we have seen that Sokal is willing to admit that non-scientific factors could have an influence, just not the primary influence on changes in scientific models. Still, Aronowitz does hit the mark in saying that, for Sokal, the scientific method of observation and reason are beyond attack. He continues

> The point [of studying social connections of science] is not to debunk science or to "deconstruct" it in order to show it is merely a fiction. This may be the postmodern project, but it is not the project of science studies. The point is to show science as a social process, to bring it down to earth, to remove the halo from its head. Scientific truth cannot be absolute; otherwise we might agree with those who have proclaimed the "end" of science. If all knowledge, including natural science, is mediated by the social and cultural context within which it has developed, then its truths are inevitably relational to the means at hand for knowing. In fact, in much of micro-physics what is called observation is often the effects of machine technologies, a reading of effects. But the reading is theory-laden. Which means pure description based on observation is not possible. Scientists require other tools such as machines, mathematics, and infer what they see from what they believe.

This is a very strong position. To claim that the use of technology, machines, must mean that "pure description based on observation is not possible" is surely an overstatement, and to say that scientists "infer what they see from what they believe" is a reactionary conclusion that, setting aside the opening disclaimer of the paragraph, does make science into a fiction. Aronowitz ends with the postmodern project that he denied a few lines before. Still, he returns to objectivity in his conclusion:

> To say that the increasing dependence of science on socially and economically permeated technology, the culture milieux within which science is done or that the political agendas of the funders invalidates results would be foolish. What it means is that scientific knowledge is not immune from broad cultural or narrow political influences and its methods cannot function as a filter. Cultural change, as much as internal debate among scientists, contributes to science – social and natural – as an evolving activity.

Aronowitz thus equates cultural change with internal debate among scientists as factors that contribute to science as an evolving activity. For Aronowitz, what we think we know depends on our cultural milieu, and reliance on the scientific method of observation and reason is no protection.

We see how far apart Sokal's two cultures really are. But the separation is not absolute: Aronowitz concludes that cultural milieu *and* debates among scientists based on the scientific method both have roles to play in the evolution of scientific models. While Aronowitz's argument does at one point reach towards the debunking of science that he attributes to "the postmodern project," he still has not answered the question *why* Lacan and others ran off with terms from mathematics and science and repurposed them. Sokal includes no fewer than eight humanists in his dossier, and each of them did use scientific terms in ways that a scientist or mathematician would not understand. If science were really so debased by postmodernism, why use the terms at all? Lacan, Kristeva, and the others discussed by Sokal used scientific terminology because of the impression it created in their work. It must not have been important for them to get the ideas exactly right from a scientific point of view, as Sokal and

Bricmont have shown us that they did not, but still they wanted the language of science and mathematics because it conveyed a different status to their work on psychoanalysis, literary criticism, and related subjects. Such language brings with it the implication of the scientific method, even if the scientific method was not used by Lacan or Kristeva in order to create their formulations.

Many humanists are not as willing as Lacan and Kristeva to associate scientific language with their work without argument and without getting it right. So, for example, Diane Larsen-Freeman and Lynne Cameron wrote *Complex Systems and Applied Linguistics* (Larsen-Freeman and Cameron 2008) to introduce complexity science to their field, which includes language development, language teaching, and discourse analysis.[1] After an initial chapter that sets the stage for the book, they offer three introductory chapters about complex systems in general, about how such systems change, and about complex systems in language. Then follow three chapters that apply a "complexity perspective" (2008: 163, and elsewhere) to topics from applied linguistics. The authors characterize this complexity perspective concisely: "Our contention, to be supported throughout this book, is that complexity theory offers applied linguistics at least an important new metaphor that brings with it new ways of thinking about issues in the field, and, maximally, may push the field towards radical theoretical change" (Larsen-Freeman and Cameron 2008: 11). The first chapter of *Complex Systems and Applied Linguistics* justifies the use of complexity science as a metaphor, and surveys related theories. The authors deserve credit for trying to educate the field about complexity science, and yet this formulation "at least an important new metaphor" gets in the way of their efforts because it is often hard to tell whether they are indeed applying a metaphor or instead are making claims about language actually having the status of a complex system. This problem turns out to be particularly severe because, as demonstrated in detail in this volume, language in use is indeed an actual complex system, while linguistic systems as we usually talk about them are only indirectly related to the operation of the complex system of speech, and so any application of complexity theory to abstract linguistic systems will remain just a metaphor. The problem, however, is not unrelated to Lacan's use of the language of topology. Was he making an analogy, a metaphor? He said not. Larsen-Freeman and Cameron claim that complex adaptive systems are "at least an important new metaphor," which leaves open the possibility, as this volume has demonstrated in a different way, that complex systems genuinely govern what happens in applied linguistics. Larsen-Freeman and Cameron are not postmodernists in the same way as Lacan and Kristeva, and yet they use the language of science with the same tension between the metaphor and objectivity as Lacan. Many humanists are attracted by the sciences, even given the difficulty to using information and language from the sciences for humanistic purposes.

[1] This section is based on Kretzschmar (2011c).

Postmodernism and speech

A few years ago I wrote a short essay called "Postmodern Dialectology" for the anniversary issue of *American Speech* (Kretzschmar 2000). It started this way:

> The American Dialect Society was founded a century ago to promote the study of dialects. It seemed self evident according to the linguistic theory of the day that "dialects," varieties of languages spoken in different regions and especially rural places (in contradistinction to the standard language of capital cities) were worth studying for what they could tell us about historical developments from their parent languages and for what they could say about the culture of their speakers. Such study also seemed urgent, since many rural dialects were believed to be disappearing as mainstream culture and education in the standard language advanced across all developed nations. I find it ironic now, at the end of the twentieth century, that many scholars are still calling for study of the same kind of dialects which were thought to be about to disappear a hundred years ago (today sometimes called "endangered dialects" in order to align such work with ecology). The truth is, as no less a scholar than William Labov has stated, that "the local accents of [American cities] are more different from each other than at any time in the past" (Labov and Ash 1997: 508). Language variation has proven more resistant in the general population to advancing standardization than anyone would have thought. While language variation carries on with gusto, it is harder and harder to characterize observed linguistic differences according to the old familiar regional terms, or even according to newer social ones... Thus the need for a label like "postmodern dialectology": what should we do when our object of study seems to be disappearing, not because of decline of language variation through the rise of language standards, but instead because we have not found what we expected to find?

Postmodernism here is not the same as what Sokal and Aronowitz claimed that it was. "Postmodern dialectology" is raised here as what must happen when "modern dialectology" has not made an adequate description of the language variation we always find around us. Figure 8.1 is a drawing from the 1920s by Russell Patterson, entitled *Where There's Smoke There's Fire.* We have some smoke from a literal glowing cigarette, but Patterson's real fire is signified here. Knowing the times then, we think immediately of newly liberated women, whether the sexuality of flappers, or the political sway of new women voters, or the growing enfranchisement of women, all of which were hot topics of the day. It was the dawn of the modern age. For many men, however, women's suffrage and empowerment represented the breakdown of the traditional order. The unladylike cigarette said as much. Modernism was all about the loss of traditional modes of life and values. Even today we can see conservatives and fundamentalists around the world struggling to hold on to that older traditional order. As Mary Klages usefully points out (2009; see also Klages 2007), modernism, in contrast to those older ways, may lament new fragmentation and incoherence but it is really about order:

> Modernity is fundamentally about order: about rationality and rationalization, creating order out of chaos. The assumption is that creating more rationality is conducive to

Figure 8.1 Russell Patterson (1893–1977), *Where There's Smoke There's Fire*, ca. 1920s (image from the Library of Congress, www.loc.gov/pictures/resource/ppmsca.01589)

creating more order, and that the more ordered a society is, the better it will function (the more rationally it will function).

Modernism seeks to reason with the world in order to make it less chaotic. Klages continues to tell us about modernism and language:

> Language, or the mode of expression used in producing and disseminating knowledge, must be rational also. To be rational, language must be transparent; it must function only to represent the real/perceivable world which the rational mind observes. There must be a firm and objective connection between the objects of perception and the words used to name them (between signifier and signified).

Language, too, must be rational and well ordered for a modernist, as in a grammar with regular, fixed rules, and as in the "firm and objective connection" between words and the things they name. Modernism, then, accepts the fact of change, like the flapper with the cigarette and all that the image implies, but seeks to apply rational, scientific means to recreate order, or at least to create the appearance of order in rational thought about a more complex reality.

Let us return now to dialects. It should be clear that the founding of the American Dialect Society and its journal was a modernist enterprise. It was a response to the perception of loss of the old regional culture, and, as time went on, dialectologists continued to address the evident fact of language variation by imagining social dialects as well as regional ones. They thus derived order from chaos by the modernist application of reason and science. The Society responded to the tenor of the times, just as Patterson's print illustrated the modern mode in the same decade that *American Speech* began publication. "Where there's smoke there's fire," as a proverbial saying, focuses our attention on the relationship between objects of perception, obvious variation in language like the visible smoke, and the ideas that these perceptions make us think of. Language variation, for the modernist, must be evidence of rational systems, regional and social dialects, that exist at a somewhat smaller scale than the overall rational system of a language.

Postmodernism, on the other hand, does not rationalize or apologize for disorder. As Klages says, "Postmodernism, in contrast [to modernism], doesn't lament the idea of fragmentation, provisionality, or incoherence, but rather celebrates that" (2009). Instead of looking for the big rational picture that confers order upon the chaos, Klages says that

> Postmodernism, in rejecting grand narratives, favors "mini-narratives," stories that explain small practices, local events, rather than large-scale universal or global concepts. Postmodern "mini-narratives" are always situational, provisional, contingent, and temporary, making no claim to universality, truth, reason, or stability.

The local and the situational are what is important for the postmodernist, not the rational explanation that pulls everything together into an ordered whole. Finally, as opposed to the modernist appreciation of science and knowledge for its own sake, Klages suggests that

> In a postmodern society, however, knowledge becomes functional – you learn things, not to know them, but to use that knowledge . . . educational policy today puts emphasis on skills and training, rather than on a vague humanist ideal of education in general. This is particularly acute for English majors. "What will you DO with your degree?"

Many postmodernists retreat from modernist rationality into parody, as in one of my favorite lines, that the best example of postmodern literature is the instruction manual for your electric toaster. Common themes in postmodern literature (here from Wikipedia) are instructive in this regard.

Irony
Playfulness, black humor
Historiographic metafiction/"faction"
Temporal distortion
Technoculture and hyperreality
Paranoia
Magic realism

Table 8.1 *Decreasing size of speech communities*

Community	Size
Kurath's (1949) regional dialects	millions
Labov's (1966) class/ethnic dialects in New York City	thousands
Milroys'(1987) social networks in Belfast	ca. 15 per network
Eckert's (2000) Jocks and Burnouts at Belden High School	central members plus others
Childs' (2005) Church Ladies, Porch Sitters in North Carolina	4 per community of practice

All of these headings undercut the big picture of rational order, by distorting time, by making history into fiction, by taking an attitude to reality through irony or paranoia, or by casting doubt on reality with technology, excessive detail, or magic. This kind of parody or subversion of great truths is what gives postmodernism a bad name for many rational souls in academia. Alan Sokal, for instance, wanted to keep a firm grip on scientific methods, on the objectivity of his sense of the external world. In this regard he is thoroughly modern. Postmodernism need be no less scientific, though, since postmodernists can also make careful, systematic observations about the world around them.

The expression of postmodernism in speech is not parodic. To paraphrase Klages, the postmodern view of language variation privileges small practices and local events, rather than large-scale universal or global concepts. With the rise of postmodernism, the scale of our language variation studies has been driven down over the decades, from large regional and urban varieties to the speech of much smaller populations (see Table 8.1). The Milroys described, not Belfast overall, but social networks in three neighborhoods in the 1980s. Penny Eckert wrote about the Jocks and Burnouts inside a high school in the 1990s, and after the turn of the century Becky Childs has written about two communities of practice in North Carolina, each with only four speakers. The locus for varieties, then, the population of the community in which we expect to find some characteristic variety of speech, has fallen in size from millions of speakers to thousands to dozens to the single digits. So, it is a postmodern view to claim, as many of us now do, that speech is essentially local, and that language variation begins in small groups at the bottom of the scale-free network of speech that rises to broad regional and social continua of speech.

So, to return to the question of what we should do when our object of study, language varieties, seems to be disappearing, not because of decline of language variation through the rise of language standards, but instead because we have not found what we expected to find? The postmodern answer is that we need to change what we expect to find from language variation. Modern dialectologists expected to find rational dialects in their changing times, and our work now is increasingly colored by postmodernism as times continue to change. Language variation has not gone away, but our way of thinking about the world has changed, again, just as it did in the modern era when the Dialect Society was founded. People who prefer to do so can still be modernists and look for the big picture. But there are good reasons to enact the postmodern acceptance of language as we find it.

First, we all form perceptions of the language around us based on our own experience with language, and what varieties each of us perceives will be somewhat different from what anybody else thinks. This is the essential finding of perceptual dialectology in the mode of Dennis Preston (e.g. 1989, 1999) and Susan Tamasi (2003): the extent of our agreement about dialects, who speaks them and what they say and what we think about them, is actually highly variable. We can see this problem of perception clearly if we think again about Patterson's flapper. Where there's smoke there's fire – but to see smoke does not just signify the burning cigarette. The smoke and the cigarette together signify the empowerment of women, and in our postmodern era each of us may think differently about that, whether we are threatened by the change, or consider the change a great improvement, or even have Patterson evoke for us a memory of the Great Depression. To return to speech, when we as postmodernists study language variation, we need to pay careful attention to our *perceptions* of language variation, not just assume with the modernists that everybody is speaking some rational subvariety of a rational, well-ordered language. The postmodern view starts small, with the individual, with each of us ourselves as an individual, and slowly builds up to larger generalizations about language in conjunction with the facts of culture that bind speakers together into communities. We can use frequency profiles effectively to help us find our way in this process, to show us how our perceptions may correspond to relative similarities in how we actually talk. If we begin with the idea that speech around us is not necessarily well-ordered and coherent, we can look, in the postmodern mode, for those aspects of it that seem to show order and coherence for particular times and situations.

The second and more important reason to choose a postmodern approach is that manual for your electric toaster: we do not just want to know intellectually about language variation, we want to do something useful with our knowledge. And I believe that we can. We just need to follow up the insight of Bill Labov (1972a), now many years ago, that black children were not trying to speak Standard English and failing, but instead were using a dialect that fit their culture. Labov is a modernist, and so he has been interested, then and now, in the big picture of African-American English. The postmodern approach is more like Walt Wolfram's, Guy Bailey's, and Tricia Cukor-Avila's findings that black speakers sound different in different communities, depending also upon their ages and social orientation towards urban life. In order to use our knowledge of language variation better, we need to focus not so much on the big picture of a language and its major dialects, but, as Klages would say, on the "situational, provisional, contingent, and temporary" settings where speakers interact with language and use the great flexibility of the complex system of speech to accomplish their communicative purposes in particular local settings and situations. Indeed, it takes a postmodern dialectologist actually to *use* knowledge of language variation effectively for our educational and commercial purposes. Big rational modernist grammars are just fine for what they are, but the big picture will not solve the particular educational problems in our local communities, or the problem of making computers understand and talk to us in the language that we actually use in our daily lives. For these functional reasons, the study of speech is no less urgent today than it was a hundred years ago, and the tenor of our times, postmodernism, tells us why and what to do about it. The study of language and complex systems is essentially a postmodern enterprise. Postmodernist linguists will do well to make careful, systematic observations of the speech in the "situational, provisional, contingent, and temporary" settings where speakers interact with language, and in so doing describe the emergence of speech patterns all around us.

When we do so, we need not abandon Sokal's reliance on the scientific method. As discussed in detail in Chapter 2, we cannot accept the model of reductionism if we are to investigate speech as a complex system. However, we need not and should not go as far as Aronowitz and undercut the value of observation of empirical evidence. It should be clear from this volume that the only way we can understand the complex system of speech is to count tokens and to assemble frequency profiles for the variants of linguistic features. The language of complex systems is not a metaphor in the study of speech; we should not apply terms like emergence or self-organization or scale-free networks to our work unless we can show the grounds for doing so. Complexity science is the right science for the study of language, and also for the study of human culture more widely. By using complex systems in our work we can build bridges from the humanities back to the natural and physical sciences, and that will be a welcome step towards the resolution of the problem of two cultures on our campuses and in our communities.

Last words

While complex systems and emergence may have originated in modern science, W. B. Yeats reflects ideas like these in *The Second Coming* (the poem is cited as this volume's epigraph). Yeats was writing at a critical moment, just after World War I had irrevocably altered the Western world. "Things fall apart" as the status quo is overwhelmed by events in a constant whirl; the connection between the falcon and the handler who controls its actions has been broken. The old settled social order has become uncertain, as the contingency of the war has provoked unpredictable change in social interactions that now proceed out of control. Yeats turns first to the spiritual for a response. His reference to the Second Coming of Christ deduces from the terrible events of the war the end of the world, for those who believe in a sequence of events determined by Christ's first incarnation. But Yeats then immediately thinks of the earlier emergence of the Sphinx in the desert, whose moment had passed when the Christian era began. And Yeats wonders what next will emerge, "its hour come round at last," in a pessimistic vision that suggests a new and darker meaning for a second coming. The poem ends with both Bethlehem and a view of world as process; for those who believe, the Second Coming will arrive at the end of time, but until then even believers have to face what comes, undetermined and unpredictable, from the whirl of human events. The Second Coming is here paired with other comings, like the Sphinx, that arise in the course of human events. These lines from Yeats capture in poetic form the sense of emergence and contingency in human culture.

We as analysts in the new humanities can use complex systems to describe emergence from human behavior and contingent events, perhaps less pessimistically than Yeats did, but with the same sense that culture is not fixed, not the inevitable product of biological causes, and instead a dynamic process from which different results arise at different times and places. It has always been possible within the traditional humanities to apply the systematic empirical observation and inductive generalizations of Aristotelian science – literary scholars have called it close reading, and those who study language have called it philology, well established and successful methods of the past in the humanities. We can use deduction, too, in the humanities, and we can be subjective, but we can also be scientific in the humanities, without reductionism and without using technological methods merely as window-dressing. Complexity is the style of explanation that fits best as a scientific approach to our subject matter and essential problems. If new humanists can locate and describe our own facts within the "scientific form" of complex systems, we have a better chance than we have had before to describe the human interactions which have always been our interest.

References

Allen, Judson. 1982. *The Ethical Poetic of the Later Middle Ages*. Toronto: University of Toronto Press.

Andor, József. 2004. The Master and His Performance: An Interview with Noam Chomsky. *Intercultural Pragmatics* 1: 93–112.

Antieau, Lamont. 2006. A distributional analysis of rural Colorado English. Diss., University of Georgia.

Aristotle. 1934. *Nicomachean Ethics*. Trans. Horace Rackham. Loeb Classical Library. Cambridge, MA: Harvard University Press.

1954. The Rhetoric. Trans W. Rhys Roberts. In *The Rhetoric and The Poetics of Aristotle*. New York: Modern Library.

1962. *Nicomachean Ethics*. Trans. Martin Ostwald. Library of the Liberal Arts. New York: Bobbs-Merrill.

1963. *Aristotle's Categories and De Interpretatione*. Trans. L. Ackrill. Oxford: Clarendon.

Aronowitz, Stanley. 1997. Alan Sokal's "Transgression." *Dissent Winter*: 107–110 (viewed online at www.physics.nyu.edu/sokal/aronowitz.html).

Arthur, W. Brian. 1999. Complexity and the Economy. *Science* 284.5411 (April 2): 107–109.

Babbitt, E. H. 1896. The English of the Lower Classes in New York City and Vicinity. *Dialect Notes* 1: 457–64.

Bailey, Guy and Natalie Maynor. 1987. Decreolization? *Language in Society* 16: 449–473.

Bailey, Richard. 2003. *Rogue Scholar: The Sinister Life and Celebrated Death of Edward H. Rulloff*. Ann Arbor, MI: University of Michigan Press.

2012. *Speaking American*. New York: Oxford University Press.

Beckner, C., Blythe, R., Bybee, J., Christiansen, M., Croft, W., Ellis, N., et al. 2009. Language is a Complex Adaptive System: Position paper. *Language Learning* 59, supplement: 1–26.

Berlin, Brent, and Paul Kay. 1969. *Basic Color Terms*. Berkeley, CA: University of California Press.

Bermudez, Jose Luis. 2010. *Cognitive Science: An Introduction to the Science of the Mind*. Cambridge: Cambridge University Press.

Biber, Douglas. 1988. *Variation across Speech and Writing*. Cambridge: Cambridge University Press.

Biber, Douglas, Susan Conrad, and Geoffrey Leech. 2002. *Longman Student Grammar of Spoken and Written English*. London: Longman.

Biber, Douglas, Susan Conrad, and Randi Reppen. 1998. *Corpus Linguistics*. Cambridge: Cambridge University Press.

Biber, Douglas, Stig Johansson, Geoffrey Leech, Susan Conrad, and Edward Finegan. 1999. *Longman Grammar of Spoken and Written English*. London: Longman.

Bloomfield, Leonard. 1933. *Language*. New York: Holt.

Bod, Rens. 1998. *Beyond Grammar: An Experience-Based Theory of Language. Center for the Study of Language and Information*. Cambridge: Cambridge University Press.

Bounds, Paulina. 2010. Perception and production of speech in the dialect of Poznań. Dissertation, University of Georgia.

Burkette, Allison. 2001. The Story of Chester Drawers. *American Speech* 76: 139–157.

2009. The Lion, the Witch, and the Armoire: Lexical Variation in Case Furniture Terms. *American Speech* 84: 315–339.

2012. Parlor Talk: Complexity from a Historical Perspective. *American Speech* 87: 391–410.

Bybee, Joan. 2001. *Phonology and Language Use*. Cambridge: Cambridge University Press.

2002. Sequentiality as the Basis for Constituent Structure. In T. Givón and B. F. Malle, eds., *The Evolution of Language out of Prelanguage*, Amsterdam: Benjamins, 109–132.

2010. *Language, Usage and Cognition*. Cambridge: Cambridge University Press.

Bybee, Joan and P. Hopper, eds. 2001. *Frequency and the Emergence of Linguistic Structure*. Amsterdam: Benjamins.

Cameron, Deborah. 2011. Evolution, Science and the Study of Literature: A Critical Response. *Language and Literature* 20: 59–72.

Carroll, John. 1956. *Language, Thought, and Reality: Selected Writings of Benjamin Whorf*. Cambridge, MA: MIT Press.

Chambers, J. K. 1996. Review of Quantitative Applications in the Social Sciences, series ed. Michael Lewis-Beck, *Journal of English Linguistics* 24: 160–64.

2003. *Sociolinguistic Theory*. 2nd edn. Oxford: Blackwell.

2009. *Sociolinguistic Theory*. Revised edn. Oxford: Wiley-Blackwell.

Chambers, J. K., and Peter Trudgill. 1998. *Dialectology*. 2nd edn. Oxford: Blackwell.

Childs, Rebecca. 2005. Investigating the local construction of identity: sociophonetic variation in Smoky Mountain African American women's speech. Dissertation, University of Georgia.

Chomsky, Noam. 1957. *Syntactic Structures*. Gravenhage: Mouton.

1965. *Aspects of the Theory of Syntax*. Cambridge, MA: MIT Press.

Coupland, Nikolas. 2011. The Sociolinguistics of Style. In R. Mesthrie, ed., *The Cambridge Handbook of Sociolinguistics*, Cambridge: Cambridge University Press, 138–156.

Croft, William. 2001. *Radical Construction Grammar: Syntactic Theory in Typological Perspective*. New York: Oxford University Press.

Dahl, Östen. 2004. *The Growth and Maintenance of Linguistic Complexity*. Amsterdam: John Benjamins.

Dinneen, Francis J. 1967. *An Introduction to General Linguistics*. Washington, DC: Georgetown University Press.

Eckert, Penelope. 2000. *Linguistic Variation as Social Practice*. Oxford: Blackwell.

Edelman, Gerald. 1987. *Neural Darwinism: The Theory of Neuronal Group Selection.* New York: Basic Books.

Ellis, Nick and Diane Larsen-Freeman, eds. 2009a. *Language as a Complex Adaptive System.* Oxford: Wiley-Blackwell. [republished in *Language Learning* 59, supplement]

2009b. Constructing a Second Language: Analyses and Computational Simulations of the Emergence of Linguistic Constructions from Usage. In Ellis and Larsen-Freeman, eds., *Language as a Complex Adaptive System.* Oxford: Wiley-Blackwell, 90–125.

Fasold, Ralph. 1990. *Sociolinguistics of Language.* Oxford: Blackwell.

Firth, J. R. 1957. *Papers in Linguistics 1934–195.* London: Oxford University Press.

Fries, Charles C. and Kenneth Pike. 1949. Coexistent Phonemic Systems. *Language* 25: 29–50.

Francis, W. Nelson and Henry Kučera. 1982. *Frequency Analysis of English Usage.* Boston, MA: Houghton Mifflin.

Gardner, Martin. 1970. Mathematical Games: The Fantastic Combinations of John Conway's New Solitaire Game "Life." *Scientific American* 223: 120–123.

Goldberg, Adele. 1995. *Constructions: A Construction Grammar Approach to Argument Structure.* Chicago, IL: University of Chicago Press.

2006. *Constructions at Work: The Nature of Generalization in Language.* Oxford: Oxford University Press.

Gottschall, Jonathan. 2008. *Literature, Science, and a New Humanities.* Basingstoke: Palgrave Macmillan.

2010. Evolutionary Literary Studies: A Foreword to Norbert Francis's Review Article, "A Modest Proposal," Being a Review of J. Gotschall, *Literature, Science, and a New Humanities. Language and Literature* 19: 301–304.

Gould, Stephen Jay. 2003. *The Hedgehog, the Fox, and the Magister's Pox: Mending the Gap Between Science and the Humanities.* New York: Three Rivers Press.

Gould, Stephen J. and Niles Eldredge. 1993. Punctuated Equilibrium Comes of Age. *Nature* 366 (18 Nov.): 222–227.

Guastello, Stephen, Matthijs Koopmans, and David Pincus, eds. 2009. *Chaos and Complexity in Psychology.* Cambridge: Cambridge University Press.

Guastello, S. and L. Liebovitch. 2009. Introduction to Nonlinear Dynamics and Complexity. In Guastello, Koopmans and Pincus, eds., *Chaos and Complexity in Psychology*, Cambridge: Cambridge University Press, 1–40.

Günther, R., L. Levitin, B. Schapiro, and P. Wagner. 1996. Zipf's Law and the Effect of Ranking on Probability Distributions. *International Journal of Theoretical Physics* 35: 395–417.

Hamilton-Brehm, Anne Marie. 2003. A foundational study of El Paso English. Diss., University of Georgia.

Hatfield, Suddarat Leerabhandh. 2005. Lexical variation of Chiangmai Dialect in Chiangmai Province in Thailand. Diss., University of Georgia.

Hawking, Stephen. 1988. *A Brief History of Time.* London: Bantam.

Hawking, Stephen, and Leonard Mlodinow. 2010. *The Grand Design.* New York: Bantam.

Hintzman, D. 1986. "Schema Abstraction" in a Multiple-Trace Memory Model. *Psychological Review* 93: 328–338.

Holland, John. 1998. *Emergence: From Chaos to Order*. New York: Basic.

Hooper, J. B. 1976. Word Frequency in Lexical Diffusion and the Source of Morphophonological Change. In W. Christie, ed., *Current Progress in Historical Linguistics*, Amsterdam: North Holland, 96–105.

Hoover, Sandra. 2001. Lexical variation and change in farming words: 1970–2001. MA thesis, University of Georgia.

Hopper, Paul. 1987. Emergent Grammar. *Berkeley Linguistics Society* 13: 139–157 (viewed at http://home.eserver.org/hopper/emergence.html).

Hopper, Paul and Elizabeth Traugott. 1993. *Grammaticalization*. 2nd edn. 2003. Cambridge: Cambridge University Press.

Horvath, B. and R. Horvath. 2001. A Multilocality Study of a Sound Change in Progress: The Case of /l/ Vocalization in New Zealand and Australian English. *Language Variation and Change* 13: 37–58.

2003. A Closer Look at the Constraint Hierarchy: Order, Contrast, and Geographical Scale. *Language Variation and Change* 15: 143–170.

Hubbell, Allan F. 1962 [1950]. *The Pronunciation of English in New York City*. New York: King's Crown Press, Columbia University. [Labov gives the date as 1962 without notice of republication or new edition, but my copy bears the date 1950. I have listed both dates to avoid any possible confusion]

Huddleston, Rodney and Geoffrey Pullum. 2002. *The Cambridge Grammar of the English Language*. Cambridge: Cambridge University Press.

Jauss, H. R. 1982. *Aesthetic Experience and Literary Hermeneutics*. Trans. Michael Shaw. Minneapolis, MN: University of Minnesota Press.

Johnson, Ellen. 1996. *Lexical Change and Variation in the Southeastern United States in the Twentieth Century*. Tuscaloosa, AL: University of Alabama Press.

Johnson, Keith. 1997. Speech Perception without Speaker Normalization. In K. Johnson and J. Mullennix, eds., *Talker Variability in Speech Processing*, San Diego, CA: Academic Press, 145–165.

Johnstone, Barbara and Scott Kiesling. 2008. Indexicality and Experience: Exploring the Meanings of /aw/-Monophthongization in Pittsburgh, *Journal of Sociolinguistics* 12: 5–33.

Jurafsky, Daniel and James Martin. 2000. *Speech and Language Processing: An Introduction to Natural Language Processing, Computational Linguistics, and Speech Recognition*. Upper Saddle River, NJ: Prentice Hall.

Kauffman, Stuart. 1995. *At Home in the Universe: The Search for the Laws of Self-Organization and Complexity*. New York: Oxford University Press.

Kenner, Hugh. 1986. Neatness Doesn't Count After All. *Discover* 7(4): 86–93.

Klages, Mary. 2007. *Literary Theory: A Guide for the Perplexed*. London: Continuum.

2009. A Summary of Postmodernism (viewed online January 2015 at www.bdavetian.com/Postmodernism.html).

Knoll, Samantha. 2009. Data from the LAMSAS project: how language varies in systematic ways. Honors thesis, University of Georgia.

Kretzschmar, William A., Jr. 1996a. Review of William Labov, *Principles of Linguistic Change*: Volume 1. *American Speech* 71: 198–205.

1996b. Quantitative Areal Analysis of Dialect Features. *Language Variation and Change* 8: 13–39.

2000. Postmodern Dialectology. *American Speech* 75: 10–12.

2003. Mapping Southern English. *American Speech* 78: 130–149.
2006. Art and Science in Computational Dialectology. *Literary and Linguistic Computing* 21: 399–410.
2008. Neural Networks and the Linguistics of Speech. *Interdisciplinary Science Reviews* 33: 336–356.
2009. *The Linguistics of Speech.* Cambridge: Cambridge University Press.
2010. Language Variation and Complex Systems. *American Speech* 85: 263–286.
2011a. The Beholder's Eye: Using Self-Organizing Maps to Understand American Dialects. In Michael Adams and Anne Curzan, eds., *Contours of English*, Ann Arbor, MI: University of Michigan Press, 53–70.
2011b. Language and Region. In R. Mesthrie, ed., *Cambridge Handbook of Sociolinguistics*, Cambridge University Press, 186–202.
2011c. Review of Larsen-Freeman and Cameron *Complex Systems and Applied Linguistics. Journal of English Linguistics* 39: 89–95.
2012. Variation in the Traditional Vowels of the Eastern States. *American Speech* 87: 378–390.
2014. Complex Systems in Aggregated Variation Analyses. In B. Szmrecsanyi and B. Wälchli, eds., *Aggregating Dialectology, Typology, and Register Analysis: Linguistic Variation in Text and Speech*, Berlin: Mouton de Gruyter, 150–173.
2015. Complex Systems and the History of the English Language. In Michael Adams, R. D. Fulk, and Laurel Brinton, eds., *Studies in the History of the English Language VI: Evidence and Method in Histories of English*, Berlin: Mouton de Gruyter.
in press. Emergence of "New Varieties" in Speech as a Complex System. In Sarah Buschfeld, Thomas Hoffmann, Magnus Huber, and Alexander Kautzsch, eds., *The Evolution of Englishes.* Amsterdam: John Benjamins.
Kretzschmar, and Irene Brockman. 2013. Scaled Measurement of Geographic and Social Speech Data. *Literary and Linguistic Computing* 28: 173–187.
Kretzschmar, William A., Jr., Jean Anderson, Joan Beal, Karen Corrigan, Lisa-Lena Opas-Hänninen, and Bartek Plichta. 2006. Collaboration on Corpora for Regional and Social Analysis. *Journal of English Linguistics* 34: 172–205.
Kretzschmar, William A., Jr., Josh Dunn, and Mi-Ran Kim. 2011. Implicational scaling in Southern speech features. Paper presented at ADS/LSA, Pittsburgh, PA.
Kretzschmar, William A., Jr., and Ellen Johnson. 1993. Using Linguistic Atlas databases for phonetic analysis. Paper presented at ACH/ALLC, Washington, DC.
Kretzschmar, Jr., William A., Ilkka Juuso, and Thomas Bailey. 2014. Computer Simulation of Dialect Feature Diffusion. *Journal of Linguistic Geography* 2: 41–57.
Kretzschmar, William A., Jr., Mi-Ran Kim, Chris Harriss, and Carl Naylor. 2006. Description of vowel formants in Dawgs' speech. Paper presented at Linguistics in the 21st Century, UGA.
Kretzschmar, William A., Jr., Sonja Lanehart, Bridget Anderson, and Becky Childs. 2007. The Relevance of Community Language Studies to HEL: The View from Roswell. In Christopher Cain and Geoffrey Russom, eds., *Managing Chaos: Strategies for Identifying Change in English, Studies in the History of the English Language*, vol. III, Berlin: Mouton de Gruyter, 173–186.
Kretzschmar, William A., Jr., Virginia G. McDavid, Theodore K. Lerud, and Ellen Johnson. 1993. *Handbook of the Linguistic Atlas of the Middle and South Atlantic States.* Chicago, IL: University of Chicago Press.

Kretzschmar, William A., Jr. and Merja Stenroos. 2012. Evidence from Surveys and Atlases in the History of the English Language. In Terttu Nevalainen and Elizabeth Traugott, eds., *The Oxford Handbook of the History of English*, Oxford: Oxford University Press, 111–122.

Kretzschmar, William A., Jr., and Susan Tamasi. 2003. Distributional Foundations for a Theory of Language Change. *World Englishes* 22: 377–401.

Kroch, Anthony. 1989. Reflexes of Grammar in Patterns of Language Change. *Language Variation and Change* 1: 199–244.

Krug, Manfred. 1998. String Frequency: A Cognitive Motivating Factor in Coalescence, Language Processing, and Linguistic Change. *Journal of English Linguistics* 26: 286–320.

Kuhl, Joseph. 2003. The idiolect, chaos and language custom far from equilibrium: conversations in Morocco. Diss., University of Georgia.

Kurath, Hans. 1949. *A Word Geography of the Eastern United States*. Ann Arbor, MI: University of Michigan Press.

Kurath, Hans and Raven I. McDavid, Jr. 1961. *The Pronunciation of English in the Atlantic States*. Ann Arbor, MI: University of Michigan Press.

Labov, William. 1963. The Social Motivation of a Sound Change. *Word* 19: 273–307.

1966. *The Social Stratification of English in New York City*. 2nd edn., 2006. Cambridge: Cambridge University Press.

1972a. The Logic of Nonstandard English. In *Language in the Inner City*, Philadelphia, PA: University of Pennsylvania Press, 201–240.

1972b. The Reflection of Social Processes in Linguistic Structures. In *Sociolinguistic Patterns*, Philadelphia, PA: University of Pennsylvania Press, 110–121.

1984. Field Methods of the Project on Linguistic Change and Variation. In J. Baugh and J. Sherzer, eds., *Language in Use*, Englewood Cliffs, NJ: Prentice Hall, 28–53.

1989. The Exact Description of the Speech Community: Short a in Philadelphia. In R. Fasold and D. Schiffrin, eds., *Language Change and Variation*, Washington, DC: Georgetown University Press, 1–57.

1994. *Principles of Linguistic Change: Internal Factors. Language in Society*, XX. Oxford: Basil Blackwell.

2001. *Principles of Linguistic Change: Social Factors. Language in Society*, XXIX. Oxford: Basil Blackwell.

2010. *Principles of Linguistic Change: Cognitive and Cultural Factors. Language in Society*, XXIX. Oxford: Wiley-Blackwell.

Labov, William and Sharon Ash. 1997. Understanding Birmingham. In C. Bernstein, T. Nunnally, and R. Sabino, eds., *Language Variety in the South Revisited*, Tuscaloosa, AL: University of Alabama Press, 508–73.

Labov, William, Charles Boberg, and Sherry Ash. 2006. *Atlas of North American English: Phonetics, Phonology and Sound Change*. Berlin: Mouton de Gruyter.

Lakoff, George and Mark Johnson. 1980. *Metaphors We Live By*. Chicago, IL: University of Chicago Press.

Langacker, Ronald. 1988. A Usage-Based Model. In Brygida Rudzka-Ostyn, ed., *Topics in Cognitive Linguistics*, Amsterdam: John Benjamins, 127–161.

Larsen-Freeman, Diane. 1997. Chaos/Complexity Science and Language Acquisition. *Applied Linguistics* 18: 141–165.

Larsen-Freeman, Diane and Lynne Cameron. 2008. *Complex Systems and Applied Linguistics.* Oxford: Oxford University Press.

Lass, Roger. 1990. How to Do Things with Junk: Exaptation in Language Evolution. *Journal of Linguistics* 26: 79–102.

Lees, Robert B. 1957. Review of Noam Chomsky, *Syntactic Structures, Language* 33: 375–408.

LePage, Robert and Andree Tabouret-Keller. 1985. *Acts of Identity.* Cambridge: Cambridge University Press.

Lewis, C. S. 1960. *Studies in Words.* Cambridge: Cambridge University Press.

Lindblom, Bjorn, Peter MacNeilage, and Michael Studdert-Kennedy. 1984. Self-Organizing Processes and the Explanation of Phonological Universals. In B. Butterworth, B. Comrie, and O. Dahl, eds., *Explanations for language universals*, New York: Mouton, 181–203.

Lindner, Tamara. 2009. The role of Cajun French in Cajun identity: opinions of South Louisiana high school students. Paper presented at SECOL, New Orleans.

Lorenz, Edward. 1972. Predictability: does the flap of a butterfly's wings in Brazil set off a tornado in Texas? Paper delivered at the American Association for the Advancement of Science, Washington.

McIntosh, Angus, Michael L. Samuels, and Michael Benskin. 1986. *A Linguistic Atlas of Late Medieval English*, Aberdeen: Aberdeen University Press.

Mandelbrot, Benoit. 1968. Information Theory and Psycholinguistics. In R. C. Oldfield and J. C. Marshall, eds., *Language: Selected Readings*, Harmondsworth: Penguin, 263–275.

1982. *The Fractal Geometry of Nature.* San Francisco, CA: Freeman.

Manning, C. and H. Schütze. 1999. *Foundations of Statistical Natural Language Processing.* Boston, MA: MIT Press.

Martinet, Andre. 1955. *Economies des changements phonetique.* Berne: Francke.

Mathesius, V. 1934. Zur synchronischen Analyse fremden Sprachguts. *Englische Studien* 70: 21–35.

Maye, Jessica, Janet Werker, and LouAnn Gerken. 2002. Infant Sensitivity to Distributional Information Can Affect Phonetic Discrimination. *Cognition* 82(3): B101–B111.

Milenkovic, Paul and Charles Read. 2000. *CSpeechSP. Software and Manual.* Madison, WI: University of Wisconsin, Dept. of Electrical and Computer Engineering, Dept. of Linguistics.

Mello, Heather. 2013. Analysis of language variation and word segmentation for a corpus of Vietnamese blogs: a sociolinguistic approach. Diss., University of Georgia.

Miller, John, and Scott Page. 2007. *Complex Adaptive Systems: An Introduction to Computational Models of Social Life.* Princeton, NJ: Princeton University Press.

Milroy, James. 1992. *Linguistic Variation and Change.* Oxford: Blackwell.

Milroy, Lesley. 1987. *Language and Social Networks.* 2nd edn. Oxford: Blackwell.

Milroy, Lesley and Matthew Gordon. 2003. *Sociolinguistics.* Oxford: Blackwell.

Milroy, James, and Lesley Milroy. 1998. *Authority in Language: Investigating Standard English.* 3rd edn. London: Routledge.

Mitchell, Melanie. 2009. *Complexity: A Guided Tour.* Oxford: Oxford University Press.

Moore, Samuel, and Thomas Knott. 1972. *The Elements of Old English.* 10th edn. Ann Arbor, MI: Wahr.

Mufwene, Salikoko. 2001. *The Ecology of Language Evolution*. Cambridge: Cambridge University Press.

2008. *Language Evolution: Contact, Competition, and Change*. London: Continuum Press.

Nerbonne, John, and Wilbert Heeringa. 2001. Computational Comparison and Classification of Dialects. *Dialectologia et Geolinguistica* 9: 69–83.

Nevalainen, Terttu and Elizabeth Traugott, eds. (2012). *The Oxford Handbook of the History of English*. Oxford: Oxford University Press.

Newman, John Henry. 1990. *The Idea of a University*. South Bend, IN: University of Notre Dame Press.

Nicolis, G. and I. Prigogine. 1977. *Self-Organization in Non-equilibrium Systems*. New York: Wiley.

Nosofsky, R. 1986. Attention, Similarity and the Identification-Categorization Relationship. *Journal of Experimental Psychology: General* 115(1): 39–57.

Paul, Herman. 1889. *Principles of the History of Language*. New York: Macmillan.

Percy, Walker. 1977. Questions They Never Asked Me. *Esquire* 88(Dec.): 170, 172, 184, 188, 190, 193–194.

Petruszewycz, M. 1973. L'histoire de la loi d'Estoup-Zipf: documents. *Mathématiques et Sciences Humaines* 44: 41–56.

Phillips, B. 1984. Word Frequency and the Actuation of Sound Change. *Language* 60: 320–342.

Pierrehumbert, Janet. 2001. Exemplar Dynamics: Word Frequency, Lenition and Contrast. In J. Bybee and P. Hopper, eds., *Frequency and the Emergence of Linguistic Structure*, Amsterdam: John Benjamins, 137–157.

Poplack, Shana. 1993. Variation Theory and Language Contact: Concepts, Methods and Data. In Dennis Preston, ed., *American Dialect Research*, Amsterdam: John Benjamins, 251–286.

Preston, Dennis. 1989. *Perceptual Dialectology*. Dordrecht: Foris.

ed. 1999. *Handbook of Perceptual Dialectology*. Amsterdam: John Benjamins.

Pulvermüller, Friedemann. 2003. *The Neuroscience of Language: On Brain Circuits of Words and Serial Order*. Cambridge: Cambridge University Press.

Quirk, Randolph, Sidney Greenbaum, Geoffry Leech, and Jan Svartvik. 1985. *A Comprehensive Grammar of the English Language*. London: Addison Wesley.

Romaine, Suzanne. 1982. *Sociolinguistic Variation in Speech Communities*. London: Arnold.

Russell, J. C. 1972. Population in Europe 500–1500. In Carlo Cipolla, ed., *The Fontana Economic History of Europe: The Middle Ages*, London: Collins/Fontana, 25–70.

Saffran, Jenny, Richard Aslin, and Elissa Newport. 1996. Statistical Learning by 8-Month-Old Infants. *Science* 274(5294): 1926–1928.

Saffran, Jenny, Elissa Newport, and Richard Aslin. 1996. Word Segmentation: The Role of Distributional Cues. *Journal of Memory and Language* 35(4): 606–621.

Sapir, Edward. 1921. *Language*. New York: Harcourt Brace.

Santa Fe Institute. www.santafe.edu

Saussure, Ferdinand de. (1916)1986. *Course in General Linguistics*. Trans Roy Harris. LaSalle, IL: Open Court.

Sawyer, R. Keith. 2005. *Social Emergence*. Cambridge: Cambridge University Press.

Schneider, Edgar. 1997. Chaos Theory as a Model for Dialect Variability and Change? In A. Thomas, ed., *Issues and Methods in Dialectology*, Bangor: University of North Wales, 122–136.

2007. *Postcolonial English: Varieties around the World*. Cambridge: Cambridge University Press.

Skousen, R. 1989. *Analogical Modeling of Language*. Dordrecht: Kluwer.

Skupin, A., and P. Agarwal. 2008. Introduction. In *Self-Organising Maps: Applications in GI Science*, London: Wiley.

Sokal, Alan. 1996. Transgressing the Boundaries: Toward a Transformative Hermeneutics of Quantum Gravity. *Social Text* 46/47: 217–252.

Sokal, Alan, and Jean Bricmont. 1998. *Fashionable Nonsense: Postmodern Intellectuals' Abuse of Science*. New York: Picador.

Stenroos, Merja, Martti Mäkinen, Simon Horobin, and Jeremy J. Smith. 2011. *The Middle English Grammar Corpus*, version 2011.1. University of Stavanger (www.uis.no/research/culture/the_middle_english_grammar_project/meg-c).

Stenroos, Merja and Kjetil Thengs. 2011. Two Staffordshires: real and typological space in the study of Late Middle English linguistic variation. Paper presented at Helsinki Corpus Festival, Helsinki.

Stockwell, Robert and Donka Minkova. 1988. The English Vowel Shift: Problems of Coherence and Explanation. In D. Kastovsky and G. Bauer, eds., *Luick Revisited: Papers Read at the Luick Symposium at Schloss Lichtenstein*, Tübingen: Gunter Narr, 355–394.

Stubbs, Michael. 2001. *Words and Phrases*. Oxford: Blackwell.

Tagliamonte, Sali. 2006. *Analysing Sociolinguistic Variation*. Cambridge: Cambridge University Press.

Thill, J., W. Kretzschmar, Jr., I. Casas, and X. Yao. 2008. Detecting Geographic Associations in English Dialect Features in North America with Self-Organising Maps. In *Self-Organising Maps: Applications in GI Science*, P. Agarwal and A. Skupin, eds., London: Wiley, 87–106.

Tomasello, Michael. 2003. *Constructing a Language*. Cambridge, MA: Harvard University Press.

Toole, John Kennedy. 1980. *A Confederacy of Dunces*. New York: Grove Press.

Trudgill, Peter. 1984. *On Dialect*. New York: New York University Press.

Upton, Clive, William A. Kretzschmar, Jr., and Rafal Konopka. 2001. *Oxford Dictionary of Pronunciation for Current English*. Oxford: Oxford University Press.

Van Geert, Paul. 1991. A Dynamic Systems Model of Cognitive Growth and Language Growth. *Psychological Review* 98: 3–53.

Van Gelder, T. 1998. The Dynamical Hypothesis in Cognitive Science. *Behavioral and Brain Sciences* 21: 615–628.

Votta, Rachel. 2007. *Communication styles in Instant Messenger use*. MA thesis, University of Georgia.

Weinreich, Uriel, William Labov, and Marvin Herzog. 1968. Empirical Foundations for a Theory of Language Change. In Winifred Lehmann and Jakob Malkiel, eds., *Directions for Historical Linguistics*, Austin, TX: University of Texas Press, 95–188.

Wilson, Edward O. 1975. *Sociobiology*. Cambridge, MA: Belknap Press.

1998. *Consilience: The Unity of Knowledge*. New York: Knopf.

Wilson, Joycelyn. 2007. Outkast'd and claimin' true: the language of schooling and education in the southern hip-hop community of practice. Diss., University of Georgia.

Wolfram, Stephen. 2002. *A New Kind of Science*. Champaign, IL: Wolfram Media.

Wright, Laura, ed. 2000. *The Development of Standard English 1300–1800*. Cambridge: Cambridge University Press.

Xu, Jiang, Stefan Kemeny, Grace Park, Carol Frattali, and Allen Braun. 2005. Language in Context: Emergent Features of Word, Sentence, and Narrative Comprehension. *NeuroImage* 25: 1002–1015.

Zelinsky, Wilbur. 1992. *The Cultural Geography of the United States*. Revised edition. Englewood Cliffs, NJ: Prentice Hall.

Zipf, George. 1949. *Human Behavior and the Principle of Least Effort*. Cambridge, MA: Addison Wesley.

Index